An Introduction to Fluid Mechanics and Transport Phenomena

FLUID MECHANICS AND ITS APPLICATIONS
Volume 86

Series Editor: R. MOREAU
MADYLAM
Ecole Nationale Supérieure d'Hydraulique de Grenoble
Boîte Postale 95
38402 Saint Martin d'Hères Cedex, France

Aims and Scope of the Series

The purpose of this series is to focus on subjects in which fluid mechanics plays a fundamental role.

As well as the more traditional applications of aeronautics, hydraulics, heat and mass transfer etc., books will be published dealing with topics which are currently in a state of rapid development, such as turbulence, suspensions and multiphase fluids, super and hypersonic flows and numerical modeling techniques.

It is a widely held view that it is the interdisciplinary subjects that will receive intense scientific attention, bringing them to the forefront of technological advancement. Fluids have the ability to transport matter and its properties as well as to transmit force, therefore fluid mechanics is a subject that is particularly open to cross fertilization with other sciences and disciplines of engineering. The subject of fluid mechanics will be highly relevant in domains such as chemical, metallurgical, biological and ecological engineering. This series is particularly open to such new multidisciplinary domains.

The median level of presentation is the first year graduate student. Some texts are monographs defining the current state of a field; others are accessible to final year undergraduates; but essentially the emphasis is on readability and clarity.

For other titles published in this series, go to
www.springer.com/series/5980

G. Hauke

An Introduction to Fluid Mechanics and Transport Phenomena

 Springer

G. Hauke
Área de Mecánica de Fluidos
Centro Politécnico Superior
Universidad de Zaragoza
C/Maria de Luna 3
50018 Zaragoza
Spain

ISBN-13: 978-90-481-7905-3
e-ISBN-13: 978-1-4020-8537-6

9 8 7 6 5 4 3 2 1

springer.com

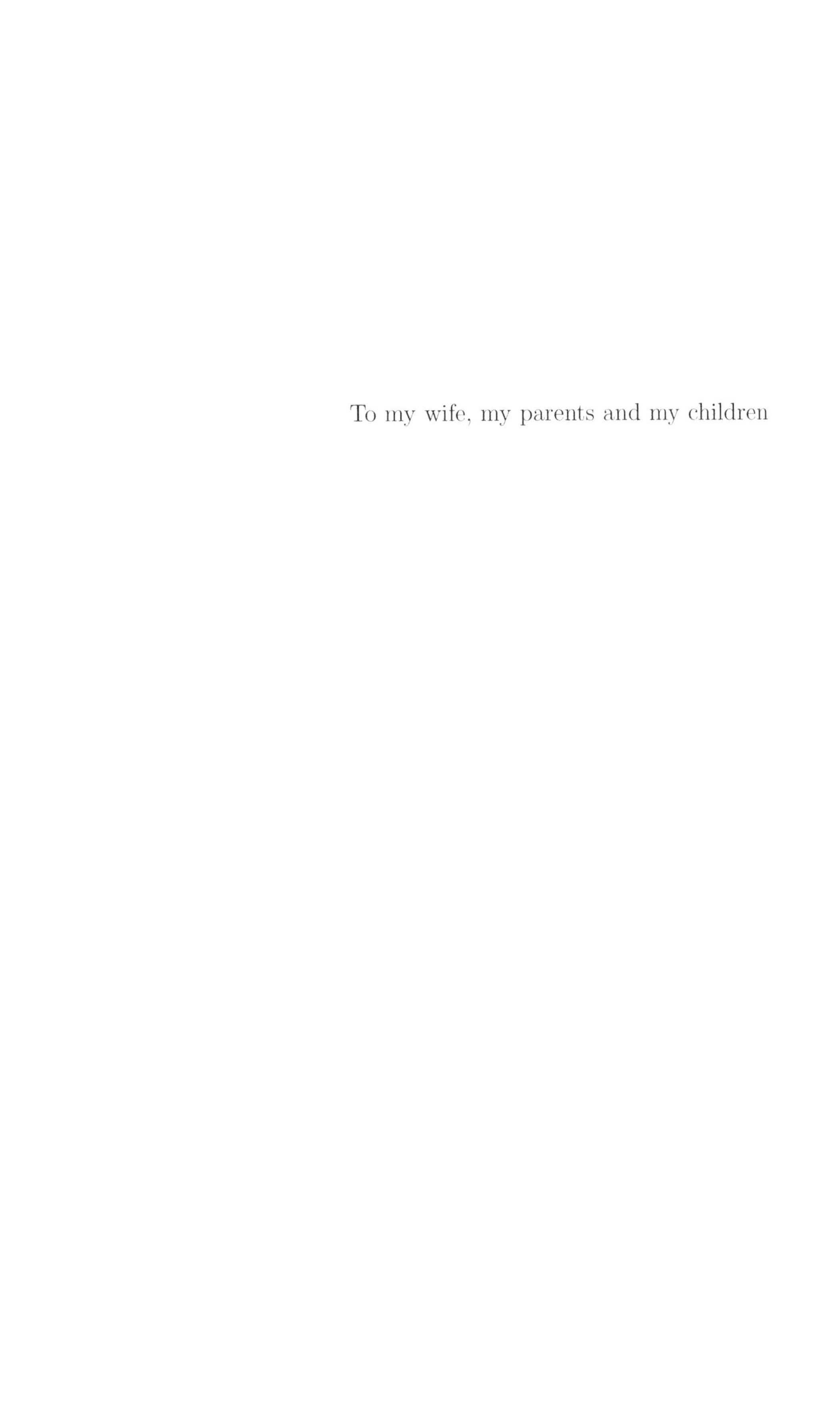

To my wife, my parents and my children

Preface

This text is a brief introduction to fundamental concepts of transport phenomena within a fluid, namely momentum, heat and mass transfer. The emphasis of the text is placed upon a basic, systematic approach from the fluid mechanics point of view, in conjunction with a unified treatment of transport phenomena.

In order to make the book useful for students, there are numerous examples. Each chapter presents a collection of proposed problems, whose solutions can be found in the Problem Solutions Appendix. Also the Self Evaluation chapter gathers exercises from exams, so readers and students can test their understanding of the subject.

Most of the content can be taught in a course of 45 hours and has been employed in the course *Transport Phenomena* in Chemical Engineering at the Centro Politécnico Superior of the University of Zaragoza. The text is aimed at beginners in the subject of transport phenomena and fluid mechanics, emphasizing the foundations of the subject.

The text is divided into four parts: Fundamentals, Conservation Principles, Dimensional Analysis;Theory and Applications, and Transport Phenomena at Interfaces.

In the first part, Fundamentals, basic notions on the subject are introduced: definition of a fluid, preliminary hypothesis for its mathematical treatment, elementary kinematics, fluid forces, especially the concept of pressure, and fluid statics.

In the Conservation Principles part, the conservation equations that govern transport phenomena are presented and explained, both in integral and differential form. Emphasis is placed on practical applications of integral equations. Also, constitutive equations for transport by diffusion are contained in this part.

In the third part, Dimensional Analysis;Theory and Applications, the important tool of dimensional analysis and the laws of similitude are explained. Also the dimensionless numbers that govern transport phenomena are derived.

The last part, Transport Phenomena at Interfaces, explains how most transport processes originate at interfaces. Some aspects of the concept of boundary layer are presented and the usage of transport coefficients to solve practical problems is introduced. Finally the analogies between transport coefficients are explained.

There are a great number of people whose help in writing this book I would like to acknowledge. First my parents, for providing an intellectually challenging environment and awakening my early interest in engineering and fluid mechanics. Professor C. Dopazo, for his inspiring passion for fluid mechanics. My family, wife and children, for their love and support. C. Pérez-Caseiras for providing ideas to strengthen the text. Many colleagues and friends, who have accompanied me during these years, especially professors T.J.R. Hughes and E. Oñate, and friends Jorge, Antonio, Connie and Ed. Finally, I would like to acknowledge the encouragement of Nathalie Jacobs. Without them, this project would not have been possible.

Zaragoza,

Guillermo Hauke
May 2008

Contents

Part IV Transport Phenomena at Interfaces

Nomenclature

Roman Symbols		Units	Dimensions
A	area	m^2	L^2
c	mixture molar concentration	mol/m^3	NL^{-3}
c_A	molar concentration of species A	mol/m^3	NL^{-3}
c_v	specific heat at constant volume	$J/(kg\ K)$	$L^2T^{-2}\Theta^{-1}$
c_p	specific heat at constant pressure	$J/(kg\ K)$	$L^2T^{-2}\Theta^{-1}$
C_D	drag coefficient	–	–
C_f	friction coefficient	–	–
D	length, diameter	m	L
D_{AB}, D_A	molecular mass diffusivity	m^2/s	L^2T^{-1}
D_v	power dissipated by viscous dissipation	W	ML^2T^{-3}
Da_I	Damköhler number	–	–
e	specific internal energy	J/kg	L^2T^{-2}
e_{tot}	specific total energy	J/kg	L^2T^{-2}
Ec	Eckert number	–	–
Eu	Euler number	–	–
$\boldsymbol{f}_m$	body force per unit mass	N/kg	LT^{-2}
$\boldsymbol{f}_s$	stress at surface	Pa	$ML^{-1}T^{-2}$
$\boldsymbol{f}_v$	body force per unit volume	N/m^3	$ML^{-2}T^{-2}$

$F, \boldsymbol{F}$	force	N	MLT^{-2}
$\boldsymbol{F}_s$	surface force	N	MLT^{-2}
$\boldsymbol{F}_v$	body force	N	MLT^{-2}
Fr	Froude number	—	—
$\boldsymbol{g}$	gravity acceleration	$\mathrm{m/s^2}$	LT^{-2}
Gr	Grashof number	—	—
h	length, depth	m	L
	heat transport coefficient	$\mathrm{W/(m^2\ K)}$	$\mathsf{MT}^{-3}\Theta$
h_m	mass transport coefficient	$\mathrm{m/s}$	LT^{-1}
H	length	m	L
$\boldsymbol{H}$	angular momentum	N m	$\mathsf{ML^2T}^{-2}$
I	surface moment of inertia	$\mathrm{m^4}$	L^4
	moment of inertia	$\mathrm{kg\ m^2}$	ML^2
$\boldsymbol{I}$	identity tensor / matrix	—	—
$\boldsymbol{j}_A$	mass flux of species A	$\mathrm{kg/(m^2\ s)}$	$\mathsf{ML^{-2}T}^{-2}$
$\boldsymbol{j}'_A$	molar flux of species A	$\mathrm{mol/(m^2\ s)}$	$\mathsf{NL^{-2}T}^{-1}$
$\boldsymbol{j}^m_A$	mass flux of species A w.r.t the molar mean velocity	$\mathrm{kg/(m^2\ s)}$	$\mathsf{ML^{-2}T}$
$\boldsymbol{j}^{m'}_A$	molar flux of species A w.r.t the molar mean velocity	$\mathrm{mol/(m^2\ s)}$	$\mathsf{NL^{-2}T}^{-1}$
J_A	mass flux of species A	$\mathrm{kg/s}$	MT^{-1}
Kn	Knudsen number	—	—
L	length, depth	m	L
Le	Lewis number	—	—
m	mass	kg	M
$\dot{m}$	mass flux	$\mathrm{kg/s}$	$\mathsf{M/T}$
$M, \boldsymbol{M}$	moment	N m	$\mathsf{ML^2/T^2}$
M	molar mass of mixture	$\mathrm{kg/kmol}$	MN^{-1}
M_A	molar mass of species A	$\mathrm{kg/kmol}$	MN^{-1}
Ma	Mach number	—	—

n_{esp}	number of chemical species in the mixture	–	–
$\boldsymbol{n}$	normal vector	–	–
Nu	Nusselt number	–	–
p	pressure	Pa	$\mathrm{ML^{-1}T^{-2}}$
$\boldsymbol{P}$	momentum	N	$\mathrm{MLT^{-1}}$
Pe	Péclet number	–	–
$\mathrm{Pe_{II}}$	Péclet II number	–	–
Pr	Prandtl number	–	–
$\boldsymbol{q}$	heat flux vector	$\mathrm{W/m^2}$	$\mathrm{MT^{-3}}$
Q	volumetric flux	$\mathrm{m^3/s}$	$\mathrm{L^3T^{-1}}$
$\dot{Q}$	heat per unit time	W	$\mathrm{ML^2T^{-3}}$
r, R	radius	m	L
$\boldsymbol{r}$	position vector	m	L
Ra	Rayleigh number	–	–
Re	Reynolds number	–	–
S	surface	$\mathrm{m^2}$	$\mathrm{L^2}$
S	Strouhal number	–	–
$S_c(t)$	control volume surface	$\mathrm{m^2}$	$\mathrm{L^2}$
$S_f(t)$	fluid volume surface	$\mathrm{m^2}$	$\mathrm{L^2}$
$\boldsymbol{S}$	deformation rate	$\mathrm{s^{-1}}$	$\mathrm{T^{-1}}$
Sc	Schmidt number	–	–
Sh	Sherwood number	–	–
St	Stanton number	–	–
t	time	s	T
T	temperature	°C or K	Θ
$\boldsymbol{u}$	velocity field	$\mathrm{m/s}$	$\mathrm{LT^{-1}}$
U	potential energy	$\mathrm{J/kg}$	$\mathrm{L^2/T^2}$
$\boldsymbol{v}$	mass average velocity	$\mathrm{m/s}$	$\mathrm{LT^{-1}}$
$\boldsymbol{v}_A$	velocity of species A	$\mathrm{m/s}$	$\mathrm{LT^{-1}}$

$\boldsymbol{v}^c$	control volume velocity	m/s	LT^{-1}
$\boldsymbol{v}^m$	molar average velocity	m/s	LT^{-1}
V	volume	m^3	L^3
	velocity	m/s	LT^{-1}
$V_c(t)$	control volume	m^3	L^3
$V_f(t)$	fluid volume	m^3	L^3
$\boldsymbol{x}$	Cartesian coordinates	m	L
	position vector	m	L
X_A	molar fraction of species A	–	–
Y_A	mass fraction of species A	–	–
$\dot{W}$	power	W	$\mathsf{ML}^2\mathsf{T}^{-3}$
We	Weber number	–	–

Greek Symbols

α	thermal diffusivity	m^2/s	$\mathsf{L}^2\mathsf{T}^{-1}$
δ	viscous boundary layer thickness	m	L
δ_T	thermal boundary layer thickness	m	L
δ_c	concentration boundary layer thickness	m	L
η_a	apparent viscosity	Pa s or kg/(m s)	$\mathsf{ML}^{-1}\mathsf{T}^{-1}$
θ	angle	rad	–
κ	thermal conductivity	W/(m K)	$\mathsf{MLT}^{-3}\Theta^{-1}$
λ	second viscosity coefficient	Pa s	$\mathsf{ML}^{-1}\mathsf{T}^{-1}$
	friction factor for pipes	–	–
	mean-free path	m	L
μ	dynamic viscosity	Pa s or kg/(m s)	$\mathsf{ML}^{-1}\mathsf{T}^{-1}$
ν	kinematic viscosity	m^2/s	$\mathsf{L}^2\mathsf{T}^{-1}$
ρ	fluid density	kg/m^3	ML^{-3}
ρ_A	mass concentration of species A	kg/m^3	ML^{-3}

σ	surface tension	N/m	MT^{-2}
	normal stress	Pa	$\mathsf{ML}^{-1}\mathsf{T}^{-2}$
$\boldsymbol{\tau},\tau$	stress tensor, stress component	Pa	$\mathsf{ML}^{-1}\mathsf{T}^{-2}$
τ'	shear stress	Pa	$\mathsf{ML}^{-1}\mathsf{T}^{-2}$
$\boldsymbol{\tau}'$	viscous stress tensor	Pa	$\mathsf{ML}^{-1}\mathsf{T}^{-2}$
ϕ_v	viscous dissipation function	W/m^3	$\mathsf{ML}^{-1}\mathsf{T}^{-3}$
ω	angular velocity	rad/s	T^{-1}
$\dot{\omega}_A$	chemical generation of species A	kg/(m^3 s)	$\mathsf{ML}^{-3}\mathsf{T}^{-1}$
$\dot{\omega}'_A$	molar chemical generation of species A	mol/(m^3 s)	$\mathsf{NL}^{-3}\mathsf{T}^{-1}$

Introduction

Most chemical processes, and the chemical and physical operations involved, imply a transport of momentum, heat and mass.

For example, let us consider a chemical reactor. The chemical compounds need to be transported into the reactor. Once in the reactor, the chemical concentrations will evolve according to the mass transport laws. In order to speed up mixing, agitation may be used to add velocity, vorticity, and turbulence to the fluid. Therefore, we are acting upon the velocity of the fluid, transferring momentum. Finally, by adding heat to the reactor, the temperature gradients generate energy transport from the heat source to the fluid particles, a process that is called heat transfer. As a consequence, in most chemical processes we can encounter mass, momentum and heat transport phenomena.

In general, the exchange of momentum, mass and energy are interrelated and appear together. For instance, mass and heat transfer are faster in the presence of agitation.

Furthermore, the laws and models that describe the transport of properties within a fluid are very similar. This is demonstrated by the existence of analogies between the three kinds of transport phenomena. Therefore, a unified study of all transport processes facilitates the learning process and deepens a relational understanding.

Finally, the chemical operations between solids, liquids and gases typically take place inside fluids (mainly liquids).

In brief, given that fluids are present in most chemical processes, it is vital for the chemical engineer to thoroughly understand fluid mechanics and transport phenomena.

Part I

Fundamentals

1

Basic Concepts in Fluid Mechanics

This chapter will define a fluid and introduce important concepts, like the continuum hypothesis and local thermodynamic equilibrium, which enable a mathematical treatment of fluid flow.

1.1 The Concept of a Fluid

In order to describe what a fluid is, two points of view are introduced: the macroscopic and the microscopic.

The macroscopic point of view consists of observing matter from the sensorial point of view: matter consists of what we touch and what we see. It is basically the engineering perspective.

In contrast, the microscopic point of view consists of describing matter through its molecular structure.

1.1.1 The Macroscopic Point of View

Experience tells us that, whereas in the solid state matter is more or less rigid, fluids are that state of matter characterized by its endless motion and deformation.

However, although the above observation is very common, a more rigorous definition is necessary. In order to formulate such a definition, let us introduce the concept of *normal* and *shear* stress.

Normal and Shear Stress

Given a surface of a body on which a force is acting, there are two types of stresses acting on that surface: normal stress and shear (or tangential) stress.

Definition 1.1 (Normal stress). *Normal stress σ is the force per unit area exerted perpendicularly to the surface over which it acts.*

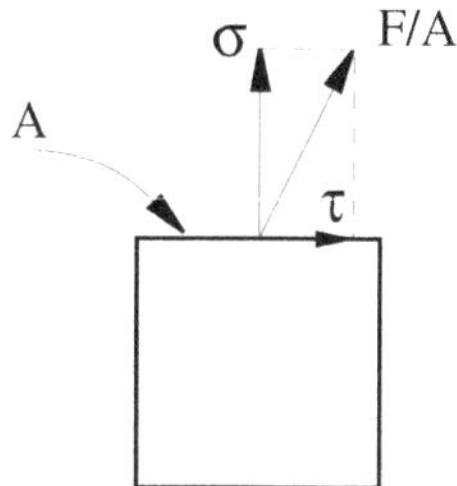

Fig. 1.1. Normal σ and shear τ stresses due to the force F acting on the surface A.

Definition 1.2 (Shear stress). *Shear (or tangential) stress τ is the force per unit area exerted tangentially to the surface over which it acts.*

Remark 1.1. The SI unit of stress is the pascal $\mathrm{Pa} = \mathrm{N/m^2}$. Since the pascal is a very small unit, in engineering applications the megapascal, $1\,\mathrm{MPa} = 10^6\,\mathrm{Pa}$, is more frequently used.

Example 1.1 (Normal and shear stress). Typical examples of normal and shear stress are, respectively, pressure and friction.

Example 1.2 (Calculation of stress components). In a horizontal plane, aligned with the x axis, there is a stress of $\boldsymbol{f}_s = (1,3)$ MPa. Calculate the normal and tangential stress.
Solution. The normal stress is the component of the stress perpendicular to the plane. In this case, the unit normal vector to the plane is $\boldsymbol{n} = (0,1)$, so the normal stress is

$$\sigma = \boldsymbol{f}_s \cdot \boldsymbol{n} = 3 \tag{1.1}$$

The tangential unit vector is $\boldsymbol{t} = (1,0)$, so the tangential projection of the stress can be calculated as

$$\tau = \boldsymbol{f}_s \cdot \boldsymbol{t} = 1 \tag{1.2}$$

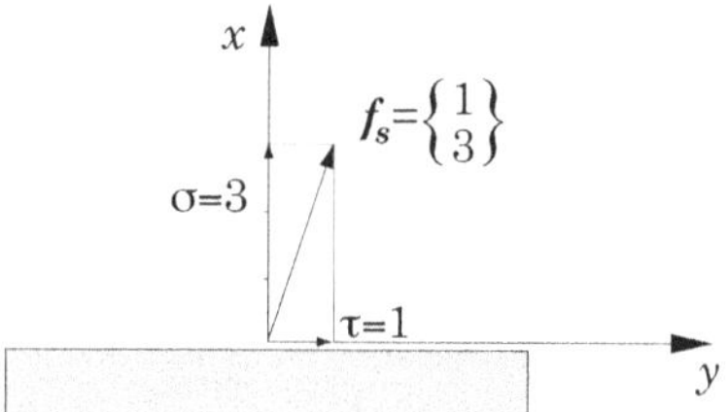

Fig. 1.2. Example 1.2. Shear and normal stresses.

Definition of a Fluid

With the concept of shear stress at hand, we can formally define a fluid. Next, two equivalent definitions of a fluid are presented.

Definition 1.3 (Fluid). *A fluid is a substance that continually deforms under the action of shear stress.*

Definition 1.4 (Fluid). *A fluid is a substance that at rest cannot withstand shear stresses.*

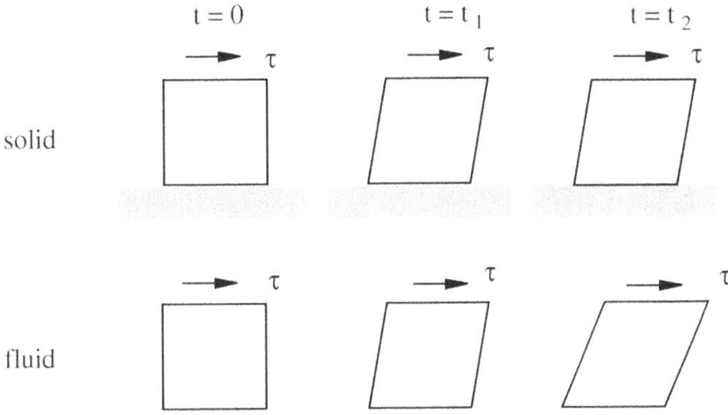

Fig. 1.3. Behavior of a small rectangular piece of solid and fluid under the action of shear stress.

Differences between Solids, Liquids and Gases

In order to further clarify what a fluid is, it is helpful to compare a fluid to a solid. In Fig. 1.3 one can observe that (below the elastic limit of deformation) a solid subject to a shear stress deforms until it reaches the equilibrium deformation, maintaining its shape thereafter. Once the force is removed, the solid recovers its original shape.

Oppositely, when subjected to a shear stress, a fluid deforms continuously until the force is relieved. A fluid does not recover its original shape when the shear stress is removed. Examples of common fluids are water, oil and air.

However, the division between solids and fluids is not always clear. There exist substances that behave like solids when the stress acts during a short period of time, but turn into fluids when the stress prolongs in time. Two examples of such substances are asphalt and the earth's crust. There are other solids that behave like fluids when the stress acting upon them reaches a threshold, like toothpaste, play-dough and melted cheese.

Definition 1.5 (Generalized fluid). *In general, a substance that for any condition obeys the definition of a fluid is called a (generalized) fluid.*

1.1.2 The Microscopic Point of View

The origin of the substances' behavior is based on their microscopic structure. Matter is formed by moving molecules, subject to various types of intermolecular forces. These forces maintain the bonding of molecules, and it is the strength of these forces that distinguishes the solid and fluid (liquid or gas) states of matter.

In a *solid* the inter-molecular forces are strong, allowing the molecules to stay at an approximately fixed position in space. For *liquids*, these cohesive forces are intermediate, weak enough to allow relative movements between molecules, but strong enough to keep the relative distance constant. Liquids, when within an open container in a gravitational field, take the shape of the container and form a surface of separation with the air, called *free surface*. Finally, in *gases* the inter-molecular forces are so weak as to allow a variable inter-molecular distance. Gases tend to expand and occupy all the available volume.

1.2 The Fluid as a Continuum

A fluid flow is characterized by the specification of *fluid variables* (sometimes also called fluid properties) such as density ρ, pressure p, temperature T, velocity vector $\boldsymbol{v}$, chemical concentration of the component A ρ_A, and so on. But according to the molecular structure of nature, if matter is made of voids and fast moving particles, how can we define each one of the above fluid variables?

For example, let us examine the fluid density. For that purpose, let us take a volume of matter δV, which will have a mass δm. In principle, the density can be calculated as the ratio between the mass δm and its volume δV,

$$\rho = \frac{\delta m}{\delta V} \tag{1.3}$$

However, depending on the size of the volume δV, we will find different values of the density.

If δV is very small, let's say microscopic, due to random molecular motion, we may at one time find one molecule, at others three, etc. Therefore, the value of the density will vary from one measurement to the next. This type of uncertainty is called *microscopic uncertainty* and is caused by the discontinuous and fluctuating nature of matter.

On the other hand, if the sampling volume is very large, such as a room, statistically speaking the number of molecules inside is going to be constant. However, due to variations of density inside the volume, the average density might differ from the actual density at the center of the room. This type of uncertainty is called *macroscopic uncertainty* and is caused by spatial variation of the fluid variables.

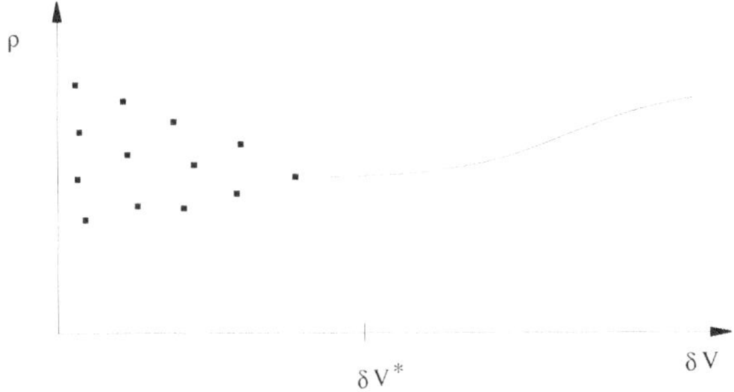

Fig. 1.4. Behavior of the measured density as a function of the sample volume.

As a consequence, in order to calculate a reasonable value of the density we need a specific size of the sampling volume δV^*, not too large nor too tiny. This volume also needs to contain enough molecules to be able to attain statistically meaningful averages. Thus, the density at a point in space is defined as

$$\rho = \lim_{\delta V \to \delta V^*} \frac{\delta m}{\delta V} \tag{1.4}$$

It has been estimated that for a stable measurement, this volume must contain around 10^6 molecules. Therefore, the size of δV^* must be $\sqrt[3]{\delta V^*}/\delta \approx 100$ where δ is the distance between molecules [14]. For instance, for air at ambient temperature, a volume δV^* of the order of 10^{-9} mm^3 contains about 3×10^7 molecules [25], a number sufficiently large to attain a correct value of density.

Furthermore, in order to be able to employ differential calculus, it will be assumed that the above definition of density yields a *continuous and continuously differentiable* function. The substances that are treated with this hypothesis are called *continuum media* and are studied in the branch of physics called continuum mechanics.

In conclusion, the *continuum hypothesis* allows us to model discontinuous matter as continuous. Certainly, every hypothesis has a range of validity. The continuum hypothesis is valid as long as a characteristic length of the flow L is much larger than that of δV^*, i.e. $\sqrt[3]{\delta V^*} << L$. This is the case for routine flows at moderate speeds, such as those encountered in chemical plants, engineering applications and vehicle aerodynamics. However, this hypothesis cannot be applied for instance to flows at pressures close to zero (called rarefied gases), such as the spacecraft reentry into the atmosphere. Also in microfluidic applications, where fluid transport phenomena take part in devices at the micro and nano scale, the continuum hypothesis can easily break down.

1.3 Local Thermodynamic Equilibrium

Fluid dynamics is tightly linked to thermodynamics, which studies the equilibrium states of substances. This equilibrium implies that the properties of matter are constant in space and time, which is rarely the case in moving fluids.

However, as in the continuum hypothesis, under certain conditions it can be assumed that each piece of fluid is in thermodynamic equilibrium. Since molecular collisions are the mechanism responsible for equilibrium, it can be assumed that there exists local thermodynamic equilibrium if within a characteristic distance for property variations $L = c/|\nabla c|$ (where c is any thermodynamic variable) there are enough collisions [8]. This condition can be expressed as $\lambda << L$, where λ is the mean free path (the path that molecules travel between collisions).

Similar arguments can be applied in time if the characteristic collision time is much smaller than the characteristic flow time scale.

Local equilibrium implies that each little piece of fluid can be considered in thermodynamic equilibrium and that at each point, the thermodynamic relations among thermodynamic properties can be safely used.

Remark 1.2. For gas flows, the applicability of the continuum hypothesis and the local thermodynamic equilibrium is usually expressed by the dimensionless *Knudsen number* [14]

$$\mathrm{Kn} = \frac{\lambda}{L} \tag{1.5}$$

For $\mathrm{Kn} < 0.01$ the medium can be considered as a continuum and the transport equations of this text are valid. For larger Kn, other models, like the Boltzmann equation, must be used.

Remark 1.3. In the case of liquids, the break-down of the continuum hypothesis manifests in anomalous diffusion mechanisms.

Problems

1.1 The stress in the plane defined by the normal $\boldsymbol{n} = \frac{1}{\sqrt{5}}(1,2)$ is $\boldsymbol{f}_s = (15, 34)$ MPa. Calculate the normal and tangential stresses in that plane.

2

Elementary Fluid Kinematics

Kinematics is the branch of fluid mechanics that studies the description of fluid motion without consideration of the forces that bring it out. There exist three formulations to describe the motion of continuum materials: the Lagrangian, the Eulerian and the arbitrary Lagrangian-Eulerian descriptions. Also, three tools to visualize fluid motion will be presented: streamlines, trajectories and streaklines. These are important in theoretical, experimental and computational fluid dynamics to visualize flow dynamics. Finally, the concept of flux will be explained.

2.1 Description of a Fluid Field

The field of transport phenomena studies the evolution of fluid variables, such as temperature, chemical concentrations, velocity and energy.

A fluid can be modeled as a numerous set of small *fluid particles* that translate, rotate and deform. This is a simple model that proves helpful to understand fluid physics. Therefore, a way to understand a fluid is to describe the motion of the particles that form the fluid.

Traditionally, there exist two ways to describe the motion of a fluid: the *Lagrangian description* and the *Eulerian description*. Presently, a combination of the above two formulations, the *Arbitrary Lagrangian-Eulerian description* (ALE), has been proved very useful in computational mechanics.

2.1.1 Lagrangian Description

This description consists of following every fluid particle, for instance, in the form of an equation for the path of each fluid particle. This approach is typically used in solid particle and rigid body mechanics and in (deformable) solid / structural mechanics. In fluid mechanics, this description is particularly suited for multiphase flows, where bubble and solid particles can be easily tracked using the Lagrangian formulation.

Mathematically, the Lagrangian description provides the position of each fluid particle $\boldsymbol{x}$ at every time instant t. Since a fluid contains an infinite number of particles, each particle is selected by specifying its position $\boldsymbol{x}_0$ at time $t = 0$,

$$\boxed{\boldsymbol{x} = \boldsymbol{x}(t, \boldsymbol{x}_0)} \tag{2.1}$$

Fig. 2.1. Lagrangian description. Position of various fluid particles as a function of time.

In this description, the acceleration of a fluid particle is determined as in kinematics of rigid bodies, where $\boldsymbol{x} = \boldsymbol{x}(t, \boldsymbol{x}_0)$ represents the position of the same particle with time. Thus,

$$\boxed{\begin{aligned} \boldsymbol{v}(t, \boldsymbol{x}_0) &= \frac{\mathrm{d}\boldsymbol{x}(t, \boldsymbol{x}_0)}{\mathrm{d}t} \\ \boldsymbol{a}(t, \boldsymbol{x}_0) &= \frac{\mathrm{d}\boldsymbol{v}(t, \boldsymbol{x}_0)}{\mathrm{d}t} = \frac{\mathrm{d}^2\boldsymbol{x}(t, \boldsymbol{x}_0)}{\mathrm{d}t^2} \end{aligned}} \tag{2.2}$$

Likewise, for other fluid properties such as evolution of temperature, for which the function

$$T = T(t, \boldsymbol{x}_0) \tag{2.3}$$

reveals the temperature T at time t of the particle that initially was at $\boldsymbol{x}_0$.

Example 2.1 (Uniform flow). The Lagrangian description of a uniform two-dimensional flow parallel to the x axis, with velocity

$$\boldsymbol{v} = \left\{ \begin{array}{c} V \\ 0 \end{array} \right\}$$

consists of the equations

$$\boldsymbol{x}(t, \boldsymbol{x}_0) = \left\{ \begin{array}{c} x_1(t, \boldsymbol{x}_0) \\ x_2(t, \boldsymbol{x}_0) \end{array} \right\} = \left\{ \begin{array}{c} x_{01} + Vt \\ x_{02} \end{array} \right\}$$

where

$$\boldsymbol{x}_0 = \left\{ \begin{array}{c} x_{01} \\ x_{02} \end{array} \right\}$$

is the position of the particle at time $t = 0$.

The acceleration is indeed zero,

$$a(t, x_0) = \frac{\mathrm{d}v(t, x_0)}{\mathrm{d}t} = \begin{Bmatrix} 0 \\ 0 \end{Bmatrix}$$

Example 2.2 (Rotating flow). In this example, the velocity field of a particle, initially at the point $(x_0, 0)$, describing a circumference centered at the origin of coordinates, with angular velocity ω is given by

$$x(t, x_0) = x_0 \begin{Bmatrix} \cos \omega t \\ \sin \omega t \end{Bmatrix}$$

In the Lagrangian formulation, the particle velocity is simply the time derivative of the position vector,

$$v = \frac{\mathrm{d}x}{\mathrm{d}t} = x_0 \omega \begin{Bmatrix} -\sin \omega t \\ \cos \omega t \end{Bmatrix}$$

and its acceleration,

$$a = \frac{\mathrm{d}v}{\mathrm{d}t} = -x_0 \omega^2 \begin{Bmatrix} \cos \omega t \\ \sin \omega t \end{Bmatrix}$$

Note that

$$a = -\omega^2 x$$

so the particle is subjected to a centripetal acceleration towards the center of rotation, of modulus $r\omega^2$.

2.1.2 Eulerian Description

Although fluid mechanics uses both, the Lagrangian and the Eulerian descriptions, the Eulerian description is the most frequently used because generally it yields simpler formulations. Indeed, in the Eulerian Description, the fluid domain is not followed as it deforms, but rather, the focus is on a fixed spatial domain, through which the fluid flows.

This formulation consists of giving the velocity field v at every spatial point x and instant of time t,

$$\boxed{v = v(x, t)} \tag{2.4}$$

Thus, this description does not provide information about the motion of each individual particle, but rather, gives information at fixed spatial points.

In this case, the acceleration of a fluid particle cannot be calculated as the partial derivative of the fluid velocity with respect to time, because $v(x, t)$

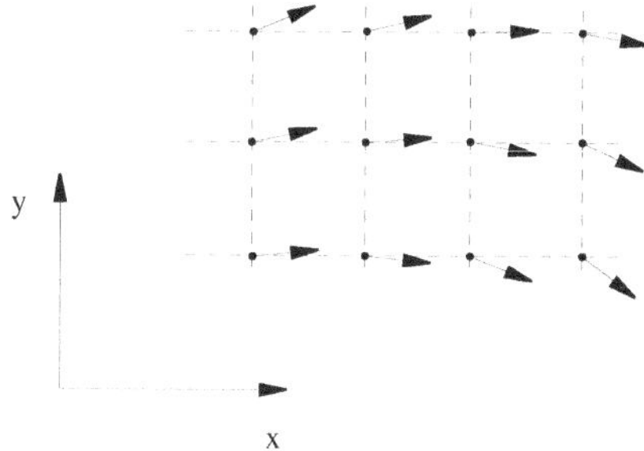

Fig. 2.2. Eulerian description. Particle velocities at fixed spatial points for a given time instant.

represents the velocity of many different particles as they travel through the same point $\boldsymbol{x}$. Therefore, to calculate the acceleration of the fluid particle, the *substantial* or *material* derivative is introduced, $D \cdot /Dt$

$$\boxed{\boldsymbol{a}(\boldsymbol{x}, t) = \frac{D\boldsymbol{v}(\boldsymbol{x}, t)}{Dt}} \tag{2.5}$$

whose definition will be given later.

Likewise, the temperature distribution is represented as the function

$$T = T(\boldsymbol{x}, t) \tag{2.6}$$

where $\boldsymbol{x}$ is the spatial coordinate and t time.

Example 2.3 (Uniform flow). The flow field of Example 2.1 in the Eulerian description would simply be

$$\boldsymbol{v}(\boldsymbol{x}, t) = \left\{ \begin{array}{c} V \\ 0 \end{array} \right\}$$

Example 2.4 (Rotating flow). In Example 2.2, the Eulerian description would not give the velocity of a single particle, but the velocity at each spatial point when different particles pass by. Thus,

$$\boldsymbol{v}(x, y, t) = \omega \left\{ \begin{array}{c} -y \\ x \end{array} \right\}$$

In this case, note that the particle acceleration is not simply the temporal derivative (which vanishes), but is given by the substantial derivative, explained below.

Example 2.5 (Measuring the water temperature in a river). Now let us present one more example to clarify the differences between the Lagrangian and Eulerian descriptions. Let us assume that we want to measure the temperature of the water in a river. If we hang a thermometer from a bridge and register the water temperature versus time at various locations on the bridge, we would be using the Eulerian description. The thermometer would be measuring the temperature of different fluid particles as they pass through a fixed point.

However, if we took a boat that was so light that it moved at the flow speed, then a thermometer glued to it would be registering the temperature of the (approximately) same fluid particle. In this case we would be using the Lagrangian description because we would be describing the temperature of the same fluid particle with time.

Obviously it is much simpler and much more accurate to fix the thermometer at a spatial point.

2.1.3 Arbitrary Lagrangian-Eulerian Description (ALE)

This description finds application in modern computational tools developed for analysis in engineering and sciences. It consists of recording data at points which move arbitrarily. For example, in a numerical computation the fluid properties are calculated at the mesh nodes. If the mesh moves, then we can use an Arbitrary Langrangian-Eulerian formulation to calculate the fluid variables at the mesh nodes.

2.2 The Substantial or Material Derivative

In classical mechanics, physical laws are formulated for a piece of matter, that is, for a particle or a set of particles. This is the case of the Lagrangian formulation, where the particle acceleration can be calculated directly as $\mathrm{d}^2 x/\mathrm{d}t^2$.

However, in the Eulerian and ALE formulations, the fluid particles are not tracked anymore. The fluid field is given as fluid properties at fixed or arbitrary points, respectively. Therefore, if evolution of the particle properties is desired, we will need specific mathematical transformations to recover the derivative *following* the fluid particle.

Let $\boldsymbol{r}(t)$ denote the position of a particle in a fluid field,

$$\boldsymbol{r}(t) = \left\{ \begin{array}{c} r_x(t) \\ r_y(t) \\ r_z(t) \end{array} \right\} \tag{2.7}$$

The velocity of this particle is

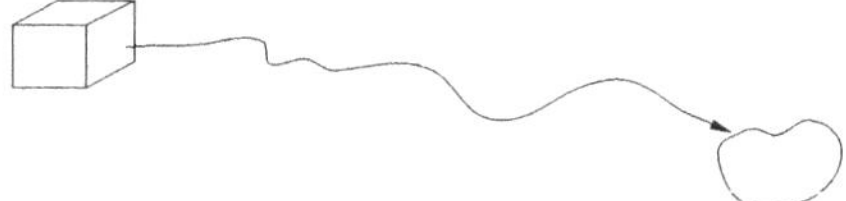

Fig. 2.3. Motion of the fluid particle.

$$\boldsymbol{v}(t) = \left\{ \begin{array}{c} u(t) \\ v(t) \\ w(t) \end{array} \right\} = \frac{\mathrm{d}\boldsymbol{r}(t)}{\mathrm{d}t} = \left\{ \begin{array}{c} \dfrac{\mathrm{d}r_x(t)}{\mathrm{d}t} \\[2mm] \dfrac{\mathrm{d}r_y(t)}{\mathrm{d}t} \\[2mm] \dfrac{\mathrm{d}r_z(t)}{\mathrm{d}t} \end{array} \right\} \tag{2.8}$$

Let a scalar Eulerian field, such as a velocity component, chemical concentration or temperature, be given by $c(x, y, z, t)$ where x, y, z are spatial coordinates. If we follow a fluid particle, the spatial coordinates are not arbitrary, but are given by the position that the fluid particle is occupying, that is, the particle trajectory, $\boldsymbol{r}(t)$. Thus, following a fluid particle,

$$c = c(r_x(t), r_y(t), r_z(t), t) \tag{2.9}$$

The total derivative of c with respect to time, Dc/Dt, which represents the *variation of c in time following the fluid particle*, can be computed by the chain rule,

$$\begin{aligned} \frac{Dc}{Dt} &= \frac{\partial c}{\partial t} + \frac{\partial c}{\partial x}\frac{\mathrm{d}r_x}{\mathrm{d}t} + \frac{\partial c}{\partial y}\frac{\mathrm{d}r_y}{\mathrm{d}t} + \frac{\partial c}{\partial z}\frac{\mathrm{d}r_z}{\mathrm{d}t} \\ &= \frac{\partial c}{\partial t} + u\frac{\partial c}{\partial x} + v\frac{\partial c}{\partial y} + w\frac{\partial c}{\partial z} \end{aligned} \tag{2.10}$$

With the help of the nabla operator $\boldsymbol{\nabla}$, which is a vector operator that computes the spatial derivatives of a function,

$$\boldsymbol{\nabla}c = \left\{ \begin{array}{c} \dfrac{\partial c}{\partial x} \\[2mm] \dfrac{\partial c}{\partial y} \\[2mm] \dfrac{\partial c}{\partial z} \end{array} \right\} \tag{2.11}$$

the derivative following the fluid particle can be expressed in tensor notation as

$$\frac{Dc}{Dt} = \frac{\partial c}{\partial t} + \left\{ \begin{array}{c} u \\ v \\ w \end{array} \right\} \cdot \left\{ \begin{array}{c} \dfrac{\partial c}{\partial x} \\[2mm] \dfrac{\partial c}{\partial y} \\[2mm] \dfrac{\partial c}{\partial z} \end{array} \right\}$$

$$= \frac{\partial c}{\partial t} + (\boldsymbol{v} \cdot \boldsymbol{\nabla})c \tag{2.12}$$

If the velocity vector is expressed in components v_i, $i = 1, 2, 3$, and the Cartesian coordinates as x_i, $i = 1, 2, 3$,

$$
\begin{aligned}
\frac{Dc}{Dt} &= \frac{\partial c}{\partial t} + u\frac{\partial c}{\partial x} + v\frac{\partial c}{\partial y} + w\frac{\partial c}{\partial z} \\
&= \frac{\partial c}{\partial t} + v_1\frac{\partial c}{\partial x_1} + v_2\frac{\partial c}{\partial x_2} + v_3\frac{\partial c}{\partial x_3} \\
&= \frac{\partial c}{\partial t} + \sum_{j=1}^{3} v_j\frac{\partial c}{\partial x_j}
\end{aligned}
\tag{2.13}
$$

Thus,

$$\boxed{\frac{Dc}{Dt} = \underbrace{\frac{\partial c}{\partial t}}_{\text{temporal}} + \underbrace{\sum_{j=1}^{3} v_j \frac{\partial c}{\partial x_j}}_{\text{convective}}} \tag{2.14}$$

This derivative is called the *substantial* or *material derivative* and represents the variation of c following a fluid particle. It is made from the *temporal term* and the *convective term*. The latter represents the transport of a property in the fluid due to its macroscopic motion.

Remark 2.1. In the last term of the above equation, the index j is repeated and, using the *Einstein summation convention* on repeated indices, the sum symbol can be eliminated (see Appendix D). Therefore, in indicial notation and for Cartesian coordinates, the substantial derivative can be written as

$$\boxed{\frac{Dc}{Dt} = \frac{\partial c}{\partial t} + v_j \frac{\partial c}{\partial x_j}} \tag{2.15}$$

If the substantial derivative is applied to the velocity vector, we obtain the acceleration of the fluid particle. In this case, the substantial derivative is applied component by component, that is, for $\boldsymbol{v} = (v_x, v_y, v_z)$

$$\boldsymbol{a} = \frac{D\boldsymbol{v}}{Dt} = \left\{ \begin{array}{c} \dfrac{Dv_x}{Dt} \\[2mm] \dfrac{Dv_y}{Dt} \\[2mm] \dfrac{Dv_z}{Dt} \end{array} \right\} \tag{2.16}$$

Definition 2.1 (Stationary or steady flow). *A fluid flow is stationary when in the Eulerian description none of the variables depends on time, i.e.,* $\frac{\partial \cdot}{\partial t} = 0$.

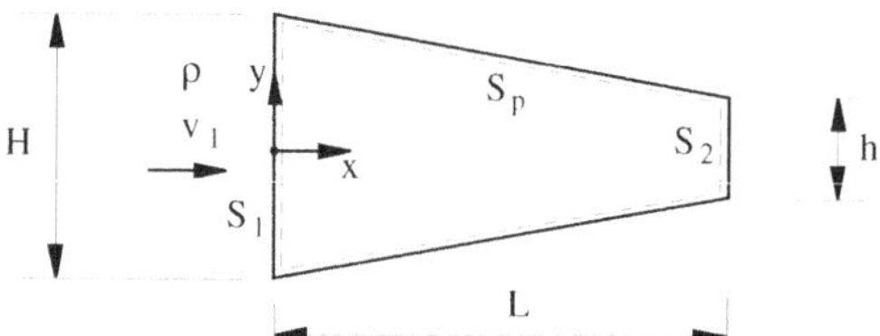

Fig. 2.4. Example: acceleration of the fluid particle inside a nozzle.

Definition 2.2 (Transient flow). *A fluid flow is said to be transient when it is not stationary.*

Example 2.6 (Flow acceleration in a converging nozzle). Let the stationary fluid flow in the nozzle of Fig. 2.4 with a decreasing cross sectional area between $x = 0$ and $x = L$ be given by the one-dimensional velocity field

$$\boldsymbol{v} = \left\{ \begin{array}{c} v_x \\ v_y \\ v_z \end{array} \right\} = \left\{ \begin{array}{c} V_0(1 + \dfrac{1}{2}\dfrac{x}{L}) \\ 0 \\ 0 \end{array} \right\}$$

Calculate the acceleration of the fluid particle.

Solution. Since the fluid flow is in the x direction, $a_y = a_z = 0$. Even though there is no temporal dependency of the flow $(\partial v_x/\partial t = 0)$ the fluid particle is still experiencing acceleration. Indeed, because the section of the nozzle is decreasing in the direction of the flow, the velocity will increase.

In order to compute the acceleration let us use the definition of the substantial derivative (2.14) applied to the component v_x of the velocity,

$$\begin{aligned}
a_x &= \cancel{\frac{\partial v_x}{\partial t}} + v_x \frac{\partial v_x}{\partial x} + v_y \cancel{\frac{\partial v_x}{\partial y}} + v_z \cancel{\frac{\partial v_x}{\partial z}} \\
&= v_x \frac{\partial v_x}{\partial x} \\
&= V_0(1 + \frac{1}{2}\frac{x}{L}) \left(\frac{V_0}{2L} \right) \\
&= \frac{V_0^2}{2L}(1 + \frac{1}{2}\frac{x}{L})
\end{aligned}$$

Example 2.7 (Rotating flow). Let us turn back to the circular motion of Example 2.2. In the Eulerian description, the velocity field would be given by

$$\boldsymbol{v}(x, y, t) = \omega \left\{ \begin{array}{c} -y \\ x \end{array} \right\}$$

And the acceleration of the fluid particle can be calculated as

$$
\begin{aligned}
\boldsymbol{a} = \frac{D\boldsymbol{v}}{Dt} &= \frac{\partial \boldsymbol{v}}{\partial t} + v_x \frac{\partial \boldsymbol{v}}{\partial x} + v_y \frac{\partial \boldsymbol{v}}{\partial y} \\
&= \begin{Bmatrix} 0 \\ 0 \end{Bmatrix} - (\omega y)\, \omega \begin{Bmatrix} 0 \\ 1 \end{Bmatrix} + (\omega x)\, \omega \begin{Bmatrix} -1 \\ 0 \end{Bmatrix} \\
&= -\omega^2 \begin{Bmatrix} x \\ y \end{Bmatrix}
\end{aligned}
$$

which coincides with the centripetal acceleration.

Remark 2.2 (Substantial Derivative in the ALE Formulation). For data points that move at the velocity $\boldsymbol{v}^{\mathrm{mesh}}$, the substantial derivative of $c(x, y, z)$ in Cartesian coordinates can be written as

$$
\frac{Dc}{Dt} = \frac{\partial c}{\partial t} + \sum_{j=1}^{3} (v_j - v_j^{\mathrm{mesh}}) \, \frac{\partial c}{\partial x_j} \tag{2.17}
$$

2.3 Mechanisms of Transport Phenomena

In a fluid there are two classes of transport phenomena: transport by convection and transport by diffusion.

The *convective transport* is due to the macroscopic fluid velocity. The fluid, with its motion, drags the fluid particles and its properties. Mathematically, the net flux by convection is modeled by the convective term of the substantial derivative,

$$
v_1 \frac{\partial c}{\partial x_1} + v_2 \frac{\partial c}{\partial x_2} + v_3 \frac{\partial c}{\partial x_3}
$$

This type of transport phenomenon is responsible, for example, for the wind transporting fallen tree leaves. At the same time as the wind blows the fallen tree leaves, it transports all the fluid properties, such as temperature, chemical concentration of the chemical species, energy, momentum and so on. In short, this transport phenomenon is caused by *fluid velocity*.

The second class of transport phenomena is due to *diffusion transport* or *molecular transport*, and it will be explained in more detail in Chapter 7 on Constitutive Equations.

Mathematically, for the case of constant coefficients, the net local balance of transport by diffusion around a fluid particle is proportional to the *diffusion coefficient* α and the Laplacian,

$$
\alpha \Delta c = \alpha \left(\frac{\partial^2 c}{\partial x_1^2} + \frac{\partial^2 c}{\partial x_2^2} + \frac{\partial^2 c}{\partial x_3^2} \right)
$$

and it contains second derivatives.

This transport phenomenon is due to the random motion (translational, vibrational, etc.) of molecules at the microscopic level, which tends to make the properties uniform. An important trait of transport by diffusion is that it can occur at a zero macroscopic velocity.

As an example, heat in a solid propagates by diffusion. In a closed room, a scent placed in a corner ends up diffusing to the whole space.

Another important characteristic of diffusive transport is that variations of fluid properties must exist. If the fluid variable is constant, then there is no transport by diffusion for that property.

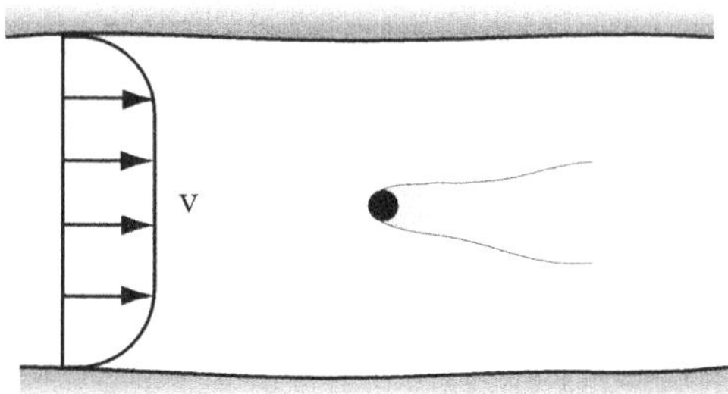

Fig. 2.5. Transport of a river spillage by convection and diffusion.

But in fluids it is common that both transport phenomena occur simultaneously. For example, let us take the spillage in the river of Fig. 2.5. The fluid velocity transports the spill downstream, and at the same time the width of the stains grows perpendicularly to the fluid velocity due to diffusion. In the absence of diffusion, the cross-sectional width of the stain would remain constant. When the flow in the river is turbulent, the mixing rate is increased considerably due to stochastic convection mechanisms, and the stain widens at an even faster rate.

2.4 Streamlines, Trajectories and Streaklines

Fluid fields can be visualized. There are three tools employed in the laboratory and in computational fluid dynamics (CFD) to visualize a velocity field: streamlines, trajectories and streaklines. They are explained next.

Definition 2.3 (Streamline). *The streamline is the line tangent at every point to the velocity vector.*

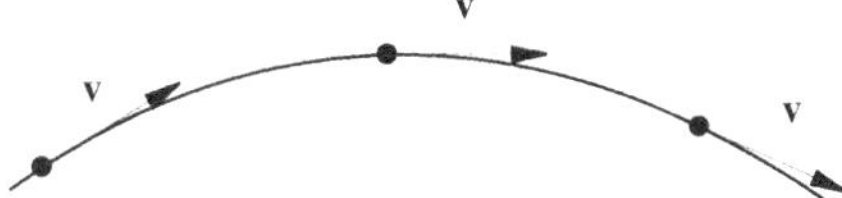

Fig. 2.6. The streamline is tangent to the velocity vector at every point.

Definition 2.4 (Trajectory). *The trajectory or path is the track followed by a fluid particle.*

Fig. 2.7. Trajectory. The fluid particle follows the plotted line.

Definition 2.5 (Streakline). *The streakline is the geometric place occupied by fluid particles that have passed by the same point at previous times.*

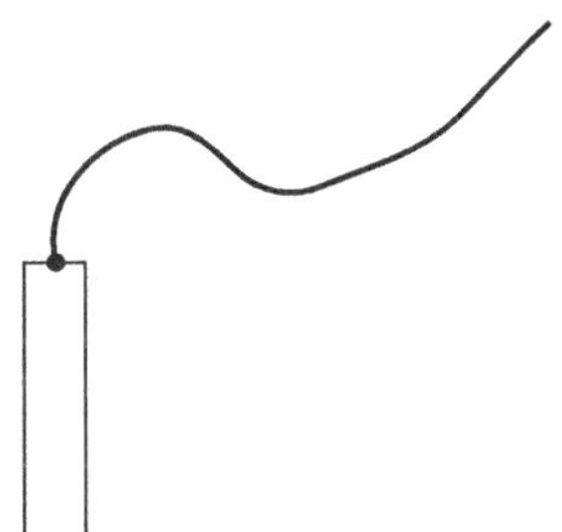

Fig. 2.8. The plume of a chimney is a streakline.

Streamlines indicate the velocity direction. They can be visualized by implanting little flags inside the fluid and observing their orientation. The streamlines can be obtained by drawing lines tangent to the flags. They are a rather mathematical object.

The trajectory is the path followed by a fluid particle. For example, the braking marks on a road indicate the position that a tire has been occupying while the wheel was being dragged. They depict, therefore, the trajectory of the tire. Another example is the path that a hiker follows to climb the peak Aneto. The same applies to fluid particles.

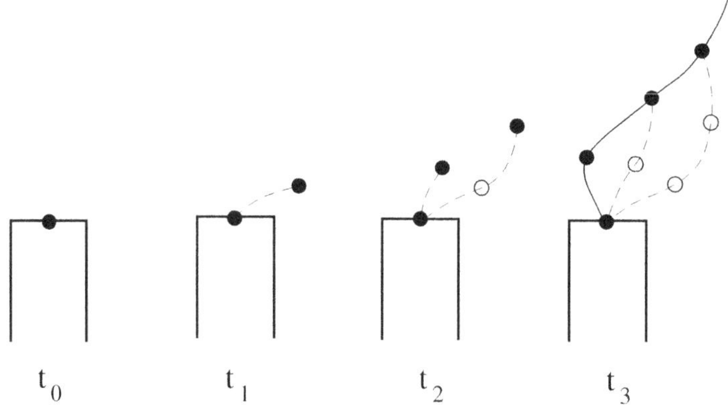

Fig. 2.9. Streakline (solid line) and trajectories (dashed lines) of the smoke particles (dots) from a chimney at successive time instants.

Finally, streaklines are the easiest element to be seen in nature or an experimental rig. Examples include a plume in the sky, the spilled colored contaminant in a river or the injected smoke in an aerodynamic tunnel. Fig. 2.9 shows how a streakline is formed and the difference between a streakline and a trajectory.

Remark 2.3. If the flow is stationary, then streamlines, trajectories and streaklines coincide.

Below, it is explained how streamlines, trajectories and streaklines are calculated. As an example, we will take the unsteady (non-stationary) two-dimensional flow field given by $u = 2x(t+1)$ and $v = 2y(t-1)$.

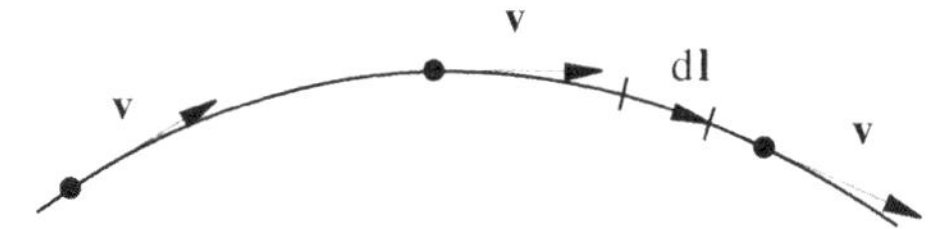

Fig. 2.10. Streamline and differential of length.

2.4.1 Calculation of Streamlines

Let $d\boldsymbol{l}$ be a differential of length along a streamline. By definition of streamline, the little piece of curve $d\boldsymbol{l}$ should be parallel to the velocity vector $\boldsymbol{v} = (u, v, w)$, that is,

$$\mathrm{d}\boldsymbol{l} \times \boldsymbol{v} = \left\{ \begin{array}{c} \mathrm{d}x \\ \mathrm{d}y \\ \mathrm{d}z \end{array} \right\} \times \left\{ \begin{array}{c} u \\ v \\ w \end{array} \right\}$$

$$= \det \begin{vmatrix} \boldsymbol{i} & \boldsymbol{j} & \boldsymbol{k} \\ \mathrm{d}x & \mathrm{d}y & \mathrm{d}z \\ u & v & w \end{vmatrix} \qquad (2.18)$$

$$= 0$$

Expanding the determinant,

$$\boxed{\frac{u}{\mathrm{d}x} = \frac{v}{\mathrm{d}y} = \frac{w}{\mathrm{d}z}} \qquad (2.19)$$

For the two-dimensional case, setting $\mathrm{d}z = 0$ and $w = 0$ in the determinant yields

$$\frac{u}{\mathrm{d}x} = \frac{v}{\mathrm{d}y} \qquad (2.20)$$

Remark 2.4. In polar coordinates, the infinitesimal lengths along the r and θ axes are $\mathrm{d}r$ and $r\mathrm{d}\theta$, respectively. Therefore, the streamlines are the solution of

$$\frac{u}{\mathrm{d}r} = \frac{v}{r\mathrm{d}\theta} \qquad (2.21)$$

with u and v the velocity components in the r and θ directions, respectively.

Example 2.8 (Streamline). Calculate the streamlines for the unsteady, two-dimensional flow field given by,

$$\begin{aligned} u &= 2x(t + 1) \\ v &= 2y(t - 1) \end{aligned}$$

Particularize for the case in which the streamline passes through the point (x_0, y_0) at all times.

Solution. Applying (2.20),

$$\frac{2x(t + 1)}{\mathrm{d}x} = \frac{2y(t - 1)}{\mathrm{d}y}$$

Integrating

$$(t + 1)\ln y = (t - 1)\ln x + \ln C$$

Thus,

$$y^{t+1} = Cx^{t-1}$$

To determine the integration constant C, the conditions of the particular case are imposed for all t,

$$y_0^{t+1} = Cx_0^{t-1}$$

and so

$$C = \frac{y_0^{t+1}}{x_0^{t-1}}$$

Finally, substituting the value of C

$$\frac{y}{y_0} = \left(\frac{x}{x_0}\right)^{\frac{t-1}{t+1}}$$

2.4.2 Calculation of Trajectories

A trajectory is the path followed by a fluid particle. Since the particle velocity is known at each spatial point, the trajectory coordinates $\boldsymbol{x}(t)$ can be obtained by integrating the equation of motion,

$$\boxed{\frac{\mathrm{d}x}{\mathrm{d}t} = u \qquad \frac{\mathrm{d}y}{\mathrm{d}t} = v \qquad \frac{\mathrm{d}z}{\mathrm{d}t} = w} \tag{2.22}$$

where (x, y, z) is the position of the particle as a function of time. As boundary condition we will need the position of a particle at a given time. Then, the variable time t can be eliminated to reach the equation of the trajectory in explicit or implicit form.

Example 2.9 (Trajectory). For the flow field of the above example, determine the trajectory of the fluid particle that passes through the point (x_0, y_0), at $t = 0$.

Solution. Integrating the equation of motion,

$$\mathrm{d}x = 2x(t + 1)\mathrm{d}t$$
$$\mathrm{d}y = 2y(t - 1)\mathrm{d}t$$

yields

$$\ln x = (t + 1)^2 + \ln C_1$$
$$\ln y = (t - 1)^2 + \ln C_2$$

Thus,

$$x = C_1 e^{(t+1)^2}$$
$$y = C_2 e^{(t-1)^2}$$

To determine the constants of integration C_1, C_2, the conditions of the problem are imposed,

$$x_0 = C_1 e^{(0+1)^2}$$
$$y_0 = C_2 e^{(0-1)^2}$$

which implies

$$C_1 \;\; = x_0/e$$

$$C_2 \;\; = y_0/e$$

Finally, the trajectory is given in parametric form through the combination of

$$\frac{x}{x_0} \;\; = e^{(t+1)^2-1}$$

$$\frac{y}{y_0} \;\; = e^{(t-1)^2-1}$$

This is a valid curve in two dimensions. Sometimes it is possible to eliminate t and write the same curve in explicit form, that is, as $y(x)$. Getting t from the first equation,

$$t \;\; = \sqrt{\ln \tfrac{x}{x_0} + 1} - 1$$

and substituting in the second one,

$$\frac{y}{y_0} \;\; = e^{\left(\sqrt{\ln \frac{x}{x_0}+1}-2\right)^2-1}$$

which is the equation of the trajectory in explicit form.

2.4.3 Calculation of Streaklines

To calculate the streaklines of a flow field, it is first necessary to compute the trajectories. The process is very similar to the one presented above, differing only in the way the boundary conditions are imposed. Assume that a tracer is injected into the flow field at the point (x_0, y_0). Then, proceed in three steps:

1. Integrate the equation of motion.
2. Calculate the integration constants, such that at time $\xi < t$ the fluid particle was at (x_0, y_0). Here ξ is the parameter that designates the particle, by the time it passed through the injection point. What we have done is to obtain all the trajectories of the particles that were injected in the flow field before the present time t.
3. Eliminate ξ.

Example 2.10 (Streakline). In the flow field of the previous example, determine the streakline that passes by x_0, y_0.
Solution. Integration of the equation of motion yields

$$x \;\; = C_1 e^{(t+1)^2}$$

$$y \;\; = C_2 e^{(t-1)^2}$$

Now, in order to determine the integration constants C_1, C_2, we search the particles that at time ξ passed by x_0, y_0:

$$x_0 = C_1 e^{(\xi+1)^2}$$

$$y_0 = C_2 e^{(\xi-1)^2}$$

yielding

$$C_1 = x_0/e^{(\xi+1)^2}$$

$$C_2 = y_0/e^{(\xi-1)^2}$$

Substituting,

$$\frac{x}{x_0} = e^{(t+1)^2-(\xi+1)^2}$$

$$\frac{y}{y_0} = e^{(t-1)^2-(\xi-1)^2}$$

The parameter ξ represents the different particles that make the streakline. For each ξ, the trajectory of a different particle is attained. As ξ is varied, we run through the various particles that make the streakline.

Getting ξ from the first equation,

$$\xi = \sqrt{(t+1)^2 - \ln \frac{x}{x_0}} - 1$$

and substituting in the second one we run through all the particles that make the streakline,

$$\frac{y}{y_0} = e^{(t-1)^2-\left(\sqrt{(t+1)^2-\ln \frac{x}{x_0}}-2\right)^2}$$

2.5 The Concept of Flux

The flux is a quantity used to measure the amount of a property transported across a surface per unit time. It is one of the most widely used concepts in fluid mechanics. As examples of daily used fluxes, we can cite the *volumetric flux* and the *mass flux*.

Before proceeding, we need to define the normal vector to a surface.

Definition 2.6 (Normal vector). *The exterior normal $\boldsymbol{n}$ at a point of a closed surface is the outward unit vector orthogonal to the surface at that point (see Fig. 2.11).*

Now we can proceed to defining the volumetric and mass flux.

Definition 2.7 (Volumetric flow rate). *The volumetric flow rate Q is the volume of fluid that crosses the surface per unit time,*

$$\boxed{Q = \int_S \boldsymbol{v} \cdot \boldsymbol{n}\, \mathrm{d}S} \tag{2.23}$$

Its dimensions are $[Q] = \mathsf{L}^3\mathsf{T}^{-1}$ and its units in the SI, m^3/s.

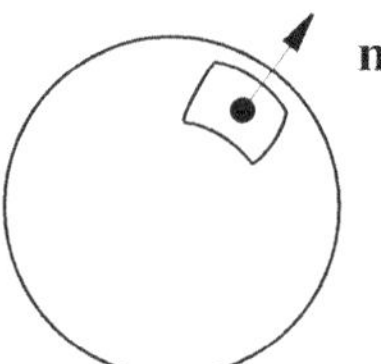

Fig. 2.11. Exterior normal to the surface of a volume.

Definition 2.8 (Mass flow rate). *The mass flux $\dot{m}$ or G is the mass per unit time that flows across a surface,*

$$\boxed{\dot{m} = \int_S \rho v \cdot n \, \mathrm{d}S}$$ (2.24)

Its dimensions are $[\dot{m}] = \mathsf{MT}^{-1}$ and its SI units, kg/s.

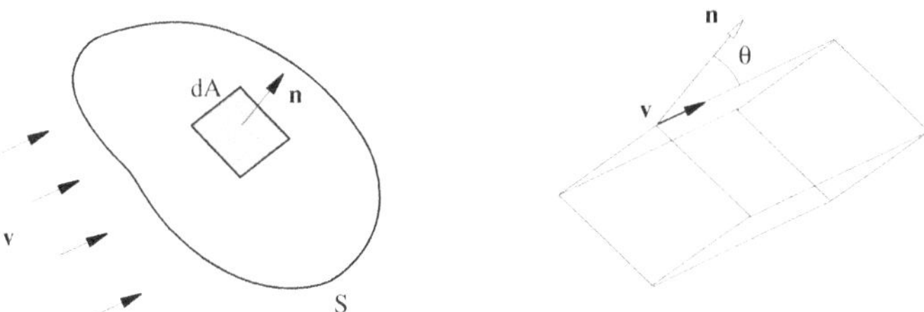

Fig. 2.12. Flux across a surface.

In order to check expression (2.23), let us take the differential of area $\mathrm{d}A$ over the surface S of Fig. 2.12. The fluid volume $\mathrm{d}\text{Vol}$ that flows during the time interval $\mathrm{d}t$ across $\mathrm{d}A$ is

$$
\begin{aligned}
\mathrm{d}\text{Vol} &= \text{base} \times \text{height} \\
&= \mathrm{d}A \times v \, \mathrm{d}t \, \cos\theta \\
&= \mathrm{d}A \, \mathrm{d}t \, (v \cdot n)
\end{aligned}
$$ (2.25)

Therefore, dividing by $\mathrm{d}t$ we calculate the volume per unit time,

$$dQ = \frac{\mathrm{d}\text{Vol}}{\mathrm{d}t} = (v \cdot n) \, \mathrm{d}A$$ (2.26)

Integrating dQ over the whole surface, the flow rate Q is calculated.

The expression for the mass flow rate is obtained following the same steps, taking into account that the differential of mass across the surface is

$$\mathrm{dmass} = \rho \, \mathrm{d}\text{Vol}$$ (2.27)

Remark 2.5. Since $\boldsymbol{n}$ is the outward normal vector to the surface, then outgoing flow rates are positive and inward flow rates, negative.

Definition 2.9 (Mean velocity). *The mean velocity $\bar{v}$ is the velocity that multiplied by the cross sectional area gives the volumetric flow rate,*

$$\boxed{\bar{v} = \frac{Q}{A}} \tag{2.28}$$

Example 2.11 (Volumetric flow rate for uniform velocity v parallel to the normal vector of the surface A). In this case, the volumetric flow rate is simply

$$Q = vA$$

The *mean velocity* is

$$\bar{v} = Q/A = v$$

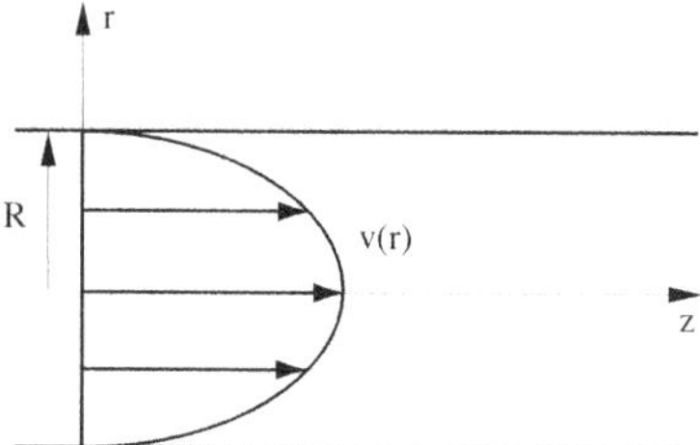

Fig. 2.13. Fully developed laminar flow in a constant section pipe.

Example 2.12 (Laminar flow in a circular cross-section pipe). The fully developed laminar axial velocity in a circular cross-section straight pipe, of radius R, obeys

$$v(r) = V_0 \left[1 - \left(\frac{r}{R} \right)^2 \right]$$

This flow is called Hagen-Poiseuille flow. Determine the volumetric flow rate in the pipe and the mean velocity.

Solution. Let us take a section perpendicular to the pipe axis, S. The volumetric flow rate is

$$Q = \int_S \boldsymbol{v} \cdot \boldsymbol{n} \, \mathrm{d}S = \int_S v(r) \, \mathrm{d}S$$

Since the velocity is constant for a given radius, we can take the surface differential $\mathrm{d}S = 2\pi r \, \mathrm{d}r$. Substituting,

$$
\begin{aligned}
Q &= \int_{r=0}^{r=R} v(r)\, 2\pi r\, \mathrm{d}r \\
&= 2\pi \int_0^R V_0 \left[1 - \left(\frac{r}{R} \right)^2 \right] r\, \mathrm{d}r \\
&= 2\pi V_0 \left[\frac{r^2}{2} - \frac{r^4}{4R^2} \right]_0^R \\
&= 2\pi V_0 \frac{R^2}{4} \\
&= \frac{\pi R^2}{2} V_0
\end{aligned}
$$

Observe that the dimensions of the above expression are correct.

The mean velocity is

$$
\bar{v} = \frac{Q}{\pi R^2} = \frac{V_0}{2}
$$

which is half the maximum velocity at the center of the pipe.

In transport phenomena, we will frequently use the *convection* or *convective* flux of a fluid property, which represents the amount of property transported across a surface per unit time. This flux is calculated from the property per unit mass ϕ. In general, we can define the convective flux of a fluid property as follows.

Definition 2.10 (Convective flux). *The convective flux of a property in a fluid with velocity v across the surface S is*

$$
\boxed{ F = \int_S \rho\phi \, \boldsymbol{v} \cdot \boldsymbol{n} \, \mathrm{d}S } \tag{2.29}
$$

where ϕ is the property per unit mass. It represents the amount of that property that crosses the surface S per unit time.

For example, for the property mass, mass per unit mass is the unity, $\phi = 1$, and the mass flow rate definition is recovered. The volumetric flux is recovered for $\phi = 1/\rho$. For the flux of internal energy, the internal energy per unit mass is $\phi = e$, where e represents the specific internal energy.

Remark 2.6. Note that for a positive $\rho\phi$, the convective flux is positive for outgoing flow ($\boldsymbol{v} \cdot \boldsymbol{n} > 0$) and negative, for incoming flow ($\boldsymbol{v} \cdot \boldsymbol{n} < 0$).

Definition 2.11 (Flux). *In general, the flux of a vector $\boldsymbol{\Phi}$ equals the integral of $\boldsymbol{\Phi} \cdot \boldsymbol{n}$ over the surface S*

$$
\boxed{ F = \int_S \boldsymbol{\Phi} \cdot \boldsymbol{n} \, \mathrm{d}S } \tag{2.30}
$$

where $\boldsymbol{n}$ is the exterior normal to the surface.

Example 2.13 (Heat flux across a surface). The heat flow or heat flux across a surface can be calculated by setting $\boldsymbol{\Phi} = \boldsymbol{q}$, where $\boldsymbol{q}$ is the vector of heat per unit surface per unit time [W/m^2]. Thus, in direction $\boldsymbol{n}$,

$$\dot{Q} = \int_S \boldsymbol{q} \cdot \boldsymbol{n} \, \mathrm{d}S \tag{2.31}$$

Table 2.1. Examples of fluxes across a surface S.

Flux	Symbol	Units	Expression
Volumetric flux	Q	m^3/s	$\int_S \boldsymbol{v} \cdot \boldsymbol{n} \, \mathrm{d}S$
Mass flux	$\dot{m}, G$	kg/s	$\int_S \rho \boldsymbol{v} \cdot \boldsymbol{n} \, \mathrm{d}S$
Flux of a property, with ϕ = prop./mass		[prop.]/s	$\int_S \rho \phi \cdot \boldsymbol{n} \, \mathrm{d}S$
Heat flux	$\dot{Q}$	W	$\int_S \boldsymbol{q} \cdot \boldsymbol{n} \, \mathrm{d}S$

Problems

2.1   Given the Eulerian fluid field

$$\boldsymbol{v}(x, y, z, t) = 3t\boldsymbol{i} + xz\boldsymbol{j} + ty^2\boldsymbol{k}$$

where $\boldsymbol{i}$, $\boldsymbol{j}$ and $\boldsymbol{k}$ are the unit vectors along the coordinate axis, determine the flow acceleration.

2.2   A fluid field is described by

$$u = \frac{x}{1+t} \; ; \; v = \frac{y}{1+2t} \; ; \; w = 0$$

where t represents time. Calculate the streamlines and trajectories that pass by x_0, y_0 at $t = 0$ and the streakline that passes by x_0, y_0 $\forall t$.

2.3   Using polar coordinates, the fluid field in a tornado can be approximated as

$$\boldsymbol{v} = -\frac{a}{r}\boldsymbol{e}_r + \frac{b}{r}\boldsymbol{e}_\theta$$

where $\boldsymbol{e}_r$ and $\boldsymbol{e}_\theta$ are the unit vectors in the directions r and θ. Show that the streamlines obey the logarithmic spiral equation

$$r = C \exp\left(-\frac{a}{b}\theta\right)$$

2.4   A two-dimensional transient velocity field is given by

$$u = 5x(1 + t) \qquad\qquad v = 5y(-1 + t)$$

where u is the x velocity component and v, the y component. Find:

(a) The trajectory $x(t)$, $y(t)$ if $x = x_0$, $y = y_0$ at $t = 0$. Is this a Lagrangian or Eulerian flow description?
(b) The streamlines that pass by x_0, y_0.
(c) The streakline that goes by x_0, y_0.

2.5   The ideal flow around a corner placed at the axis origin is given by

$$\begin{aligned} u_x &= ax \\ u_y &= -ay \end{aligned}$$

with a a constant. Determine the streamlines and draw the streamline that goes by the point $(1, 1)$ indicating the flow direction for $a > 0$. Calculate the substantial derivative.

2.6   The velocity field in an irrotational vortex, like the ones present in cyclones, is given by

$$\begin{aligned} u_x &= -Ky/(x^2 + y^2) \\ u_y &= Kx/(x^2 + y^2) \end{aligned}$$

Determine the streamlines and draw a few of them.

2.7   Check if the velocity field of the above exercise can be expressed in polar coordinates as

$$\begin{aligned} u_r &= 0 \\ u_\theta &= K/r \end{aligned}$$

and calculate the streamlines in polar coordinates.

2.8   Calculate the volumetric and mass flow rate, Q and $\dot{m}$, respectively, across the slit S of width w shown in the Figure, where the velocity vector has an angle of 60° with the x axis and its magnitude is given by $v = v_0(b^2 - x^2)/b^2$. The fluid density is ρ.

2.9   In film condensation along a vertical plate in a vapor atmosphere (see the Figure), Nusselt found out that in laminar flow the velocity profile at a station x is

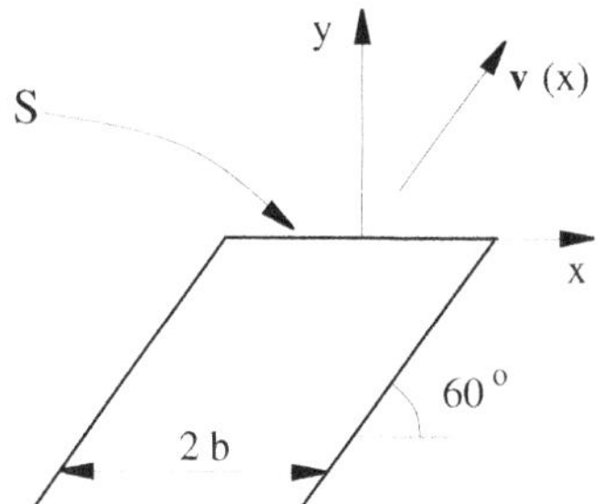

Problem 2.8. Slit geometry.

$$v_x(y) = \frac{(\rho_l - \rho_v)g\delta^2}{\mu} \left[\frac{y}{\delta} - \frac{1}{2}\left(\frac{y}{\delta}\right)^2 \right] \qquad 0 \le y \le \delta(x)$$

where ρ_l and ρ_v are the density of the fluid in the liquid and vapor phase, respectively, and μ the liquid density. Find the volumetric flow rate per unit width at any value of x.

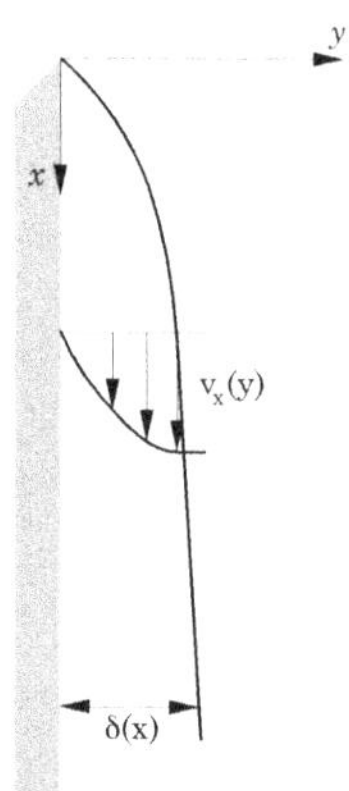

Problem 2.9. Film condensation on a vertical plate.

2.10   Write the expression of the kinetic energy flux across a surface S.

3

Fluid Forces

The intensity of transport processes crucially depends on fluid motion. This is one of the salient features of transport phenomena in fluids. And since the fluid motion is impelled by the forces applied to the fluid, these fluid forces are of paramount importance in determination of the mentioned processes. Therefore, this chapter introduces the main forces that for engineering purposes act upon a fluid, namely, body forces, surface forces and surface tension. The concept of static pressure is also explained.

3.1 Introduction

Eventually, we are interested in describing transport phenomena in moving fluids. In classical mechanics, a motion is described by the second law of Newton. For example, the acceleration a of a particle of mass m subject to external forces $\sum F$ is given by

$$\sum F = ma$$

Therefore, to determine the motion of the fluid, we need to know the forces that act upon the fluid particles.

Typically, in fluid engineering applications three types of forces are considered:

(a) Body forces
(b) Surface forces
(c) Line forces or surface tension

They are explained next.

3.2 Body Forces

This type of force acts upon the whole material volume at a distance, without contact. An example is gravitational force. This force field can be defined in two different ways.

The first way is as a volume force, which is defined as the exerted force per unit volume

$$\boldsymbol{f}_v = \frac{\text{force}}{\text{volume}}$$

It can also be defined as a massic force, the force exerted over the body per unit mass,

$$\boldsymbol{f}_m = \frac{\text{force}}{\text{mass}}$$

The differential force $\mathrm{d}\boldsymbol{F}_v$ acting on a differential element of volume $\mathrm{d}V$ and mass $\mathrm{d}m = \rho\,\mathrm{d}V$ can be calculated as

$$\mathrm{d}\boldsymbol{F}_v = \boldsymbol{f}_v\,\mathrm{d}V \tag{3.1}$$

$$\mathrm{d}\boldsymbol{F}_v = \boldsymbol{f}_m\,\mathrm{d}m = \boldsymbol{f}_m\,\rho\,\mathrm{d}V \tag{3.2}$$

The total force over a given domain V can be calculated by integration

$$\boxed{\begin{aligned} \boldsymbol{F}_v &= \int_V \boldsymbol{f}_v\,\mathrm{d}V \\[2ex] \boldsymbol{F}_v &= \int_V \rho\,\boldsymbol{f}_m\,\mathrm{d}V \end{aligned}} \tag{3.3}$$

Note that both types of forces (per unit mass or volume) can be used to compute the total force $\boldsymbol{F}_v$ exerted over a body. As a consequence,

$$\boxed{\boldsymbol{f}_v = \rho\,\boldsymbol{f}_m} \tag{3.4}$$

As examples of this type of force, we can cite

(a) The gravity

$$\boldsymbol{f}_v = \rho\boldsymbol{g}$$

where ρ is the fluid density (ML^{-3}) and $\boldsymbol{g}$, the acceleration of the gravitational field (with dimensions LT^{-2} and $|\boldsymbol{g}| = 9.81$ m/s^2).

(b) The electromagnetic force

$$\boldsymbol{f}_v = \rho_e \boldsymbol{E} + \boldsymbol{J} \times \boldsymbol{B}$$

(c) Inertial forces due to the acceleration of relative systems of reference

$$\boldsymbol{f}_m = -(\boldsymbol{a}_0 + \dot{\boldsymbol{\Omega}} \times \boldsymbol{r} + \boldsymbol{\Omega} \times (\boldsymbol{\Omega} \times \boldsymbol{r}) + 2\boldsymbol{\Omega} \times \boldsymbol{v})$$

where $\boldsymbol{a}_0$ is the acceleration of the origin of the relative system; $\boldsymbol{\Omega}$ is the angular velocity of the reference system; $\dot{\boldsymbol{\Omega}}$, its angular acceleration; $\boldsymbol{r}$, the position vector and $\boldsymbol{v}$, the fluid velocity with respect to the relative system of reference.

3.3 Surface Forces

As the name indicates these forces act upon the surface of the fluid particle or upon the surface of the considered fluid domain.

These forces are computed from *stresses* (see Chapter 1),

$$f_s = \frac{\text{force}}{\text{surface}}$$

The force acting on a differential element of surface dS is

$$dF_s = f_s \, dS \tag{3.5}$$

whose integration over the given surface S yields the global force

$$\boxed{F_s = \int_S f_s \, dS} \tag{3.6}$$

Examples of stresses that produce surface forces are

(a) The pressure p, where the stress acting on the direction n is given by

$$f_s = -pn$$

(b) The friction force.

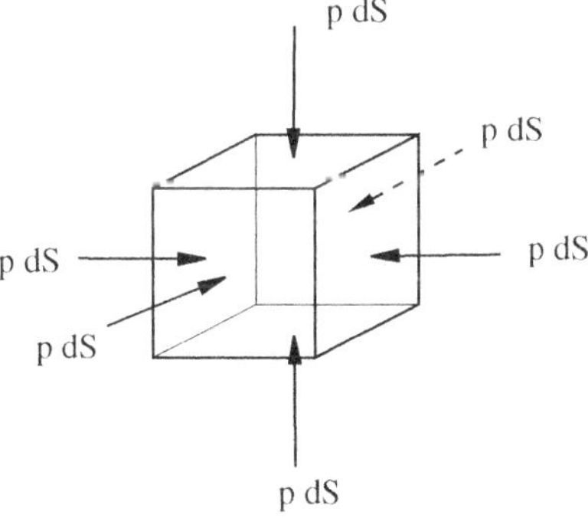

Fig. 3.1. Pressure forces acting on the fluid particle.

3.3.1 The Stress Tensor

The mathematical formulation to calculate surface stresses (and forces) is much more complex than that of the body forces because they depend on the orientation of the surface. Indeed, at a point in space there are infinitely many planes, and this needs to be accounted for.

The fundamental idea is that to calculate $\boldsymbol{f}_s$ in any plane, the stresses acting on three perpendicular planes are necessary. That is, we need 3 planes $\times$ $3\frac{\text{forces}}{\text{plane}} = 9$ data per point. These stresses are gathered in the stress tensor.

Definition 3.1 (Stress tensor). *The stress tensor is the matrix $\boldsymbol{\tau}$ whose components τ_{ij} represent stresses, i.e., ratios*

$$stress = \frac{force}{surface}$$

For three-dimensional flows, $1 \leq i,j \leq 3$, and

$$\boldsymbol{\tau} = \begin{bmatrix} \tau_{11} & \tau_{12} & \tau_{13} \\ \tau_{21} & \tau_{22} & \tau_{23} \\ \tau_{31} & \tau_{32} & \tau_{33} \end{bmatrix}$$

Definition 3.2 (Components of the stress tensor). *The component τ_{ij} of the stress tensor is the stress that acts on the plane perpendicular to the axis i and in the direction of the axis j for foreground faces and in the opposite direction for background faces.*

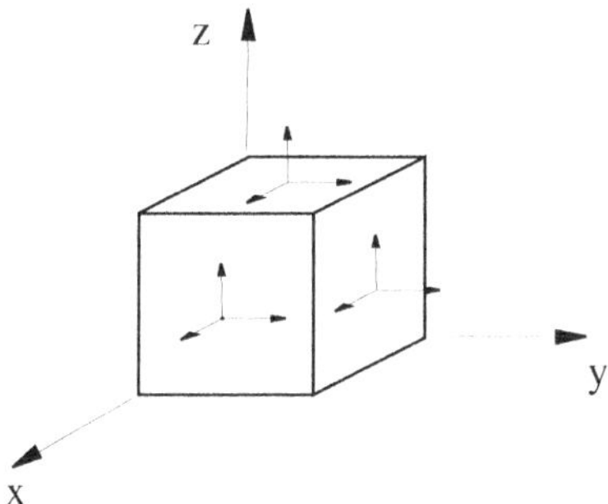

Fig. 3.2. Positive stresses that act upon the foreground faces of an elemental cube.

Remark 3.1. Foreground faces are those where the normal vector is aligned to a coordinate axis and background faces those where the normal vector is in the opposite direction to a coordinate axis.

Remark 3.2. The opposite convention for i and j is equally found. This has no practical implication since, as shown below, the stress tensor is symmetric.

Definition 3.3 (Normal stress). *Normal stresses are those perpendicular to the face upon which they act, i.e.* τ_{11}, τ_{22} *and* τ_{33}.

Definition 3.4 (Shear or tangential stress). *Shear or tangential stresses are those tangent to the surface upon which they act, i.e.* τ_{12}, τ_{13}, τ_{21}, τ_{23}, τ_{31} *and* τ_{32}.

Remark 3.3. In Cartesian coordinates, the components of the stress tensor are also denoted by τ_{xx}, τ_{yy}, τ_{zz}, τ_{xy}, τ_{xz}, and τ_{yz}.

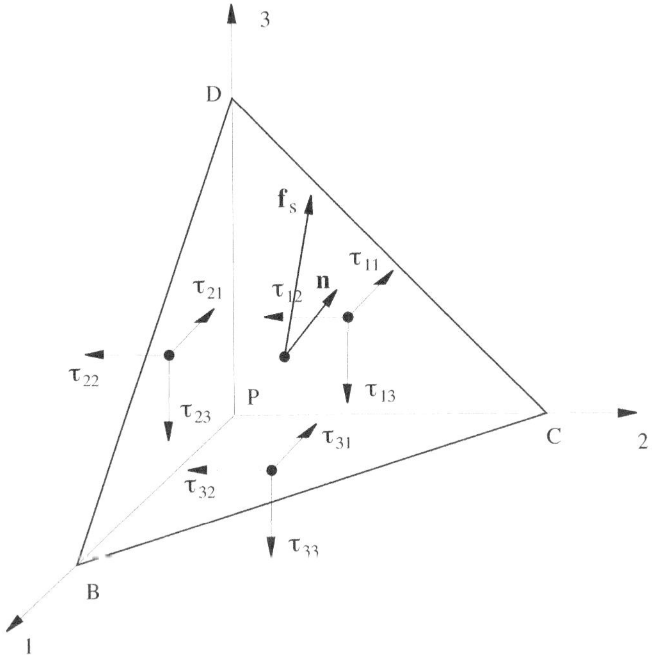

Fig. 3.3. Infinitesimal tetrahedron employed to obtain the stress tensor at the point P.

Derivation of the Stress Tensor

In order to determine the general expression of the stress at a point P from the stresses on three perpendicular planes, let us select the infinitesimal fluid volume of Fig. 3.3 and apply the second Newton's law

$$\sum \boldsymbol{F} = m\boldsymbol{a}$$

The following notation is employed for the vectors and areas,

$$PB = \mathrm{d}x_1$$
$$PC = \mathrm{d}x_2$$
$$PD = \mathrm{d}x_3$$

$$\mathrm{d}A = BCD$$
$$\mathrm{d}A_1 = CPD = n_1\mathrm{d}A$$
$$\mathrm{d}A_2 = BPD = n_2\mathrm{d}A$$
$$\mathrm{d}A_3 = BPC = n_3\mathrm{d}A$$

$$\boldsymbol{f}_s = \begin{Bmatrix} f_{s1} \\ f_{s2} \\ f_{s3} \end{Bmatrix}$$
$$\boldsymbol{n} = \begin{Bmatrix} n_1 \\ n_2 \\ n_3 \end{Bmatrix}$$
$$\boldsymbol{a} = \begin{Bmatrix} a_1 \\ a_2 \\ a_3 \end{Bmatrix}$$
$$\boldsymbol{f}_v = \rho\boldsymbol{g} = \rho \begin{Bmatrix} g_1 \\ g_2 \\ g_3 \end{Bmatrix}$$

The infinitesimal element is a tetrahedron of volume

$$\mathrm{d}V = \frac{1}{3}\left(\frac{1}{2}\mathrm{d}x_1\mathrm{d}x_2\right)\mathrm{d}x_3 = \frac{1}{6}\mathrm{d}x_1\mathrm{d}x_2\mathrm{d}x_3$$

and mass

$$\mathrm{d}m = \rho\mathrm{d}V$$

The first component of the equation of motion is

$$f_{s1}\mathrm{d}A - \tau_{11}n_1\mathrm{d}A - \tau_{21}n_2\mathrm{d}A - \tau_{31}n_3\mathrm{d}A + \rho g_1\frac{1}{6}\mathrm{d}x_1\mathrm{d}x_2\mathrm{d}x_3 = \rho a_1\frac{1}{6}\mathrm{d}x_1\mathrm{d}x_2\mathrm{d}x_3$$

Ignoring the infinitesimals of third order $\mathrm{d}x_1\mathrm{d}x_2\mathrm{d}x_3$ compared to the infinitesimals of second order $\mathrm{d}A$,

$$f_{s1} - \tau_{11}n_1 - \tau_{21}n_2 - \tau_{31}n_3 = 0$$

Likewise, for the directions 2 and 3 we have, respectively,

$$f_{s2} - \tau_{12}n_1 - \tau_{22}n_2 - \tau_{32}n_3 = 0$$

$$f_{s3} - \tau_{13}n_1 - \tau_{23}n_2 - \tau_{33}n_3 = 0$$

Summarizing,

$$\begin{Bmatrix} f_{s1} \\ f_{s2} \\ f_{s3} \end{Bmatrix} = \begin{bmatrix} \tau_{11} & \tau_{21} & \tau_{31} \\ \tau_{12} & \tau_{22} & \tau_{32} \\ \tau_{13} & \tau_{23} & \tau_{33} \end{bmatrix} \begin{Bmatrix} n_1 \\ n_2 \\ n_3 \end{Bmatrix}$$

As will be shown, the stress tensor is *symmetric* and, therefore, in tensor notation we can write,

$$\boxed{\boldsymbol{f}_s = \boldsymbol{\tau}\,\boldsymbol{n}} \tag{3.7}$$

Conclusion. If $\boldsymbol{\tau}$ is known, the surface force acting on any direction can be calculated.

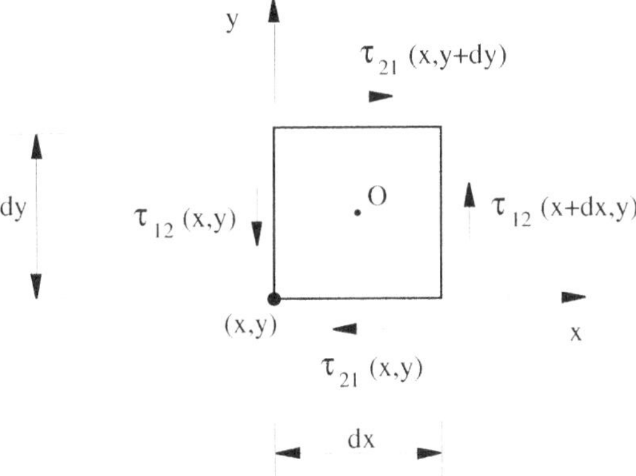

Fig. 3.4. Stresses on a cube acting on the torque balance around the z axis.

Symmetry of the Stress Tensor

The tensor $\boldsymbol{\tau}$ is symmetric, i.e.

$$\boxed{\begin{aligned} \tau_{12} &= \tau_{21} \\ \tau_{13} &= \tau_{31} \\ \tau_{23} &= \tau_{32} \end{aligned}} \tag{3.8}$$

Proof. Let us take an infinitesimal cube of sides dx, dy, dz (see Fig. 3.4). The angular momentum equation with respect to an axis parallel to z that passes through the point O implies

$$\sum M_O = I\dot{\Omega}$$

with $\sum M_O$ the external moments around the z axis acting on the cube, I the moment of inertia and $\dot{\Omega}$ the angular acceleration around the z axis. Fig. 3.4 shows all the components of the stress tensor that contribute to the above moment balance.

As in the previous section, the external moment is an infinitesimal of third order while the moment of inertia is an infinitesimal of fourth order. Therefore, in the $dV \rightarrow 0$ limit, the above equation tends to

$$\sum M_O = 0$$

Taking moments with respect to the point O, and using Taylor series,

$$\sum M_O = \left(\tau_{12} + \frac{\partial \tau_{12}}{\partial x}\mathrm{d}x\right)\frac{\mathrm{d}x}{2}(\mathrm{d}y\mathrm{d}z) + \tau_{12}\frac{\mathrm{d}x}{2}(\mathrm{d}y\mathrm{d}z)$$
$$- \left(\tau_{21} + \frac{\partial \tau_{21}}{\partial y}\mathrm{d}y\right)\frac{\mathrm{d}y}{2}(\mathrm{d}x\mathrm{d}z) - \tau_{21}\frac{\mathrm{d}y}{2}(\mathrm{d}x\mathrm{d}z) = 0$$

Ignoring the fourth-order infinitesimals against the third-order ones yields

$$\tau_{12}\mathrm{d}x\mathrm{d}y\mathrm{d}z - \tau_{21}\mathrm{d}x\mathrm{d}y\mathrm{d}z = 0$$

and since $\mathrm{d}x\mathrm{d}y\mathrm{d}z \neq 0$,

$$\tau_{12} = \tau_{21}$$

Similarly for the rest of the stresses.

3.3.2 The Concept of Pressure

In order to introduce the concept of pressure, let us take a fluid at rest. According to the definition of a fluid, a fluid at rest *cannot withstand any tangential stresses*, implying that all the off-diagonal components of the stress tensor vanish,

$$\tau_{12} = \tau_{13} = \tau_{23} = 0$$

and the stress tensor reduces to normal stresses,

$$\boldsymbol{\tau} = \begin{bmatrix} \tau_{11} & 0 & 0 \\ 0 & \tau_{22} & 0 \\ 0 & 0 & \tau_{33} \end{bmatrix} \tag{3.9}$$

Let us take the rectangular prism of Fig. 3.5. The equilibrium of forces in the x direction implies

$$-\tau_{xx}\mathrm{d}z\mathrm{d}y + \tau\mathrm{d}s\mathrm{d}y\,\sin\theta = 0$$

Taking into account that from the geometry of the prism

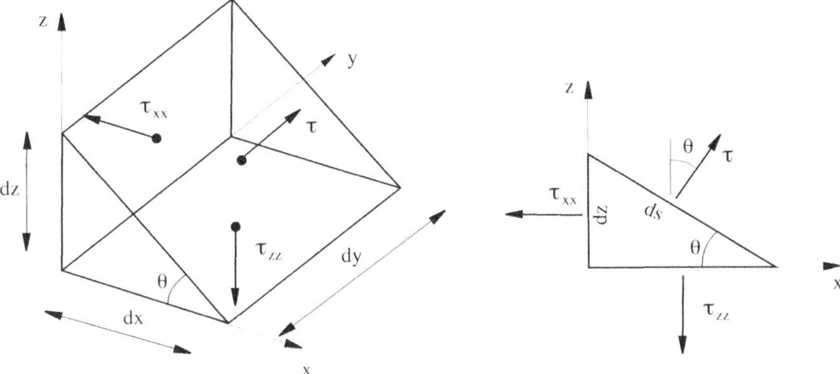

Fig. 3.5. Infinitesimal volume to derive the concept of pressure.

$$\frac{\mathrm{d}z}{\mathrm{d}s} = \sin\theta$$

we conclude

$$\boxed{\tau_{xx} = \tau \qquad \text{for any plane } \theta}$$

Likewise, it holds

$$\tau_{xx} = \tau_{zz}$$

Consequence. In a fluid at rest, the normal stresses are identical in any spatial direction. This normal stress is called *pressure*, $p > 0$, and because it is a compression stress, it is negative, so that

$$\tau = \tau_{xx} = \tau_{yy} = \tau_{zz} = -p$$

Therefore, the *stress tensor in a fluid at rest* is:

$$\boxed{\tau = \begin{bmatrix} -p & 0 & 0 \\ 0 & -p & 0 \\ 0 & 0 & -p \end{bmatrix}} \tag{3.10}$$

Note that the pressure is a *scalar* and it acts equally in any spatial direction. The direction of the force that the pressure causes is determined by the normal to the surface $\boldsymbol{n}$. The minus sign indicates that pressure is a negative normal stress, also called *compression*, that acts in the opposite direction to the exterior normal. When the normal stress is positive, then it acts in the direction of the exterior normal, producing *traction*.

Remark 3.4. This type of tensor, proportional to the identity tensor, is called an *isotropic tensor*.

Remark 3.5. For a fluid in motion, the stress tensor is the sum of the action due to the pressure plus a contribution from the motion, called the *viscous stress tensor* $\boldsymbol{\tau}'$, which is introduced in Chapter 7.

Remark 3.6. For a liquid, the pressure is a mechanical variable. It can also be defined as the average normal stress of the fluid particle, technically speaking, the *trace* of the stress tensor.

Remark 3.7. For a gas, the pressure is a thermodynamic variable. In this case, the pressure cannot be defined as the average normal stress of the fluid particle.

Units of Pressure

As a normal stress, the units of pressure are equivalent to

$$\frac{\text{force}}{\text{surface}}$$

See Table 3.1.

Table 3.1. SI and non-SI common units of pressure.

Name	Symbol	Equivalence
pascal (SI)	Pa	$1\ \mathrm{Pa} = 1\ \mathrm{N/m^2}$
bar	bar	$1\ \mathrm{bar} = 10^5\ \mathrm{Pa}$
atmosphere (standard)	atm	$1\ \mathrm{atm} = 101\,300\ \mathrm{Pa}$
meter of water	mH$_2$O	$1\ \mathrm{mH_2O} = 1\,000g\ \mathrm{Pa}$
millimeter of mercury	mmHg	$1\ \mathrm{mmHg} = 13\,600g\ \mathrm{Pa}$
kilogram-force per square centimeter (TS)	kgf/cm^2	$1\ \mathrm{kgf/cm^2} = 10^4\,g\ \mathrm{Pa}$

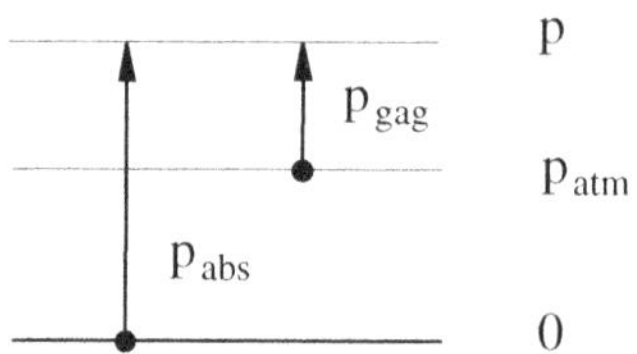

Fig. 3.6. Absolute and gage pressure.

Absolute and Gage Pressure

The absolute pressure p_{abs} represents the force per unit area exerted by the molecules against the walls. It is always positive and in the case of a gas, it corresponds to the thermodynamic pressure.

The gage pressure p_{gag} is the pressure measured relatively to the atmospheric pressure p_{atm},

$$\boxed{p_{\text{gag}} = p_{\text{abs}} - p_{\text{atm}}} \tag{3.11}$$

It can be positive or negative, its minimum value being $-p_{\text{atm}}$ (see Fig. 3.6).

The atmospheric pressure is taken as

$$p_{\text{atm}} = 101.3 \text{ kPa} = 1.013 \text{ bar}$$

3.4 Surface Tension

At interfaces between two substances, the inter-molecular forces at both sides differ, appearing to be an additional force. At the macroscopic level, the interfacial forces can be modeled by the *surface tension* σ,

$$\text{surface tension} = \frac{\text{force}}{\text{length}}$$

which causes a force tangent to the interface and orthogonal to any line through the interface, of modulus

$$\boxed{\mathrm{d}F_l = \sigma \, \mathrm{d}l} \tag{3.12}$$

The surface tension depends on the pair of substances that form the interface and on the temperature. When the surface tension is positive, the molecules of each phase tend to be repelled back to their own phase. This is the case, for instance, of two *inmiscible* liquids. When the surface tension is negative, the molecules of both phases tend to mix, like two miscible liquids. In the case of a liquid/gas interface, the surface tension tends to maintain the interface (or free surface) straight.

An important situation appears when three substances meet forming three interfaces, for instance, at a wall/liquid/gas interface. In this case, the line, which is the intersection of the three interfaces, is called the *contact line*. The angle that two interfaces form at the contact line is called the *contact angle* and depends on the surface tension of all interfaces. Therefore, the contact angle depends solely on the three substances and the temperature.

The dimensionless number that controls the significance of the surface tension force is called the *Weber* number,

$$\boxed{\text{We} = \quad = \frac{\rho U^2 L}{\sigma} \qquad\qquad \text{Weber number}}$$

Table 3.2. Surface tension for various pairs of fluids at $20°$ in N/m $\times\ 10^3$ [8].

	Water	Mercury	Ethanol	Benzene
Air	72.8	487	22	29
Water		375	< 0	35

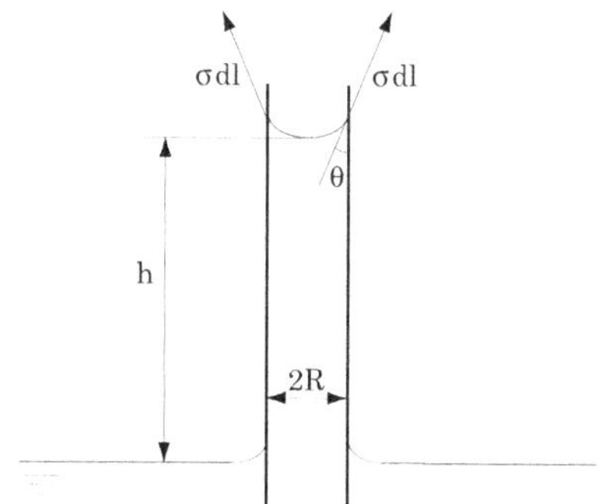

Fig. 3.7. Liquid ascending through a capillary.

where U and L are a characteristic velocity and length scale of the problem, respectively.

Example 3.1 (Capillary forces). Calculate the maximum height that a liquid can ascend through a capillary.

Solution. Due to surface tension, a liquid inside a capillary is subject to an ascending force of $2\pi R\sigma \cos\theta$, where θ is the contact angle between the water and the solid surface of the capillary. Since the weight of the liquid column $\rho g \pi R^2 h$ is equal to the ascending force, the maximum height that a liquid can climb through a capillary due to surface tension is

$$h = \frac{2\sigma \cos\theta}{\rho g R}$$

Example 3.2 (Pressure jump across a bubble. Young-Laplace equation). Calculate the pressure jump across a spherical bubble in a fluid at rest.

Solution. Let us cut the bubble into two hemispheres and invoke force equilibrium. In each of the hemispheres, the pressure difference causes a force equal to

$$(p_{\mathrm{in}} - p_{\mathrm{out}})\pi R^2$$

whereas the surface tension creates a force equal to

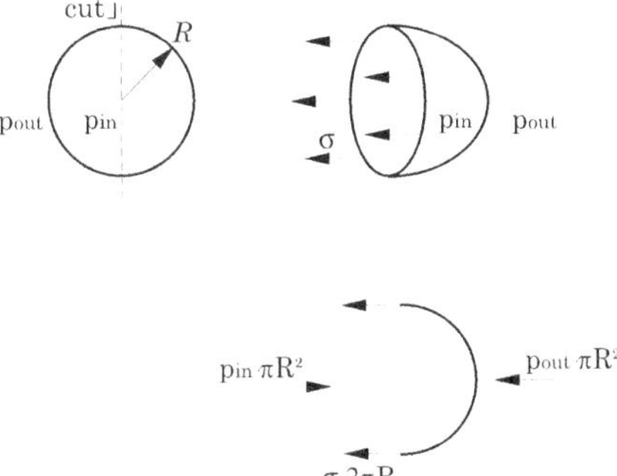

Fig. 3.8. Pressure jump across a bubble immersed in a fluid.

$$\sigma 2\pi R$$

Combining

$$p_{\text{in}} - p_{\text{out}} = \frac{2\sigma}{R}$$

3.5 Summary

Table 3.3 summarizes the variables which are employed to calculate the three types of forces of most interest in engineering applications.

Table 3.3. Summary of variables to calculate forces acting on a fluid.

Force			Calculated from	Units	Dimensions
Body force	$\boldsymbol{F}_v$	$\boldsymbol{f}_v$	volumic force	N/m^3	$ML^{-2}T^{-2}$
		$\boldsymbol{f}_m$	massic force	N/kg	LT^{-2}
Surface force	$\boldsymbol{F}_s$	$\boldsymbol{f}_s$	stress/traction	$N/m^2 = Pa$	$ML^{-1}T^{-2}$
		τ	stress tensor	$N/m^2 = Pa$	$ML^{-1}T^{-2}$
Line force	$\boldsymbol{F}_l$	σ	surface tension	N/m	MT^{-2}

Problems

3.1 Determine the normal stress acting on the plane of the Figure, where the non-vanishing stress tensor components are: $\tau_{xx} = 35$ kgf/cm², $\tau_{yy} = -7$ kgf/cm² and $\tau_{xy} = \tau_{yx} = 2$ kgf/cm².

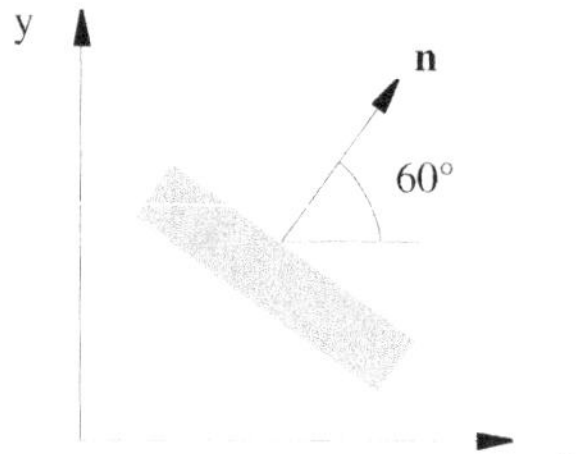

Problem 3.1. Stresses over an inclined plane.

3.2 Prove that the gage pressure cannot be lower than $-p_{\mathrm{atm}}$.

3.3 As shown in the Figure, to experimentally determine the surface tension of a gas/liquid interface, the Du Noüy balance measures the force to detach a thin ring from a liquid double meniscus. Ignoring the curvature of the meniscus, determine the surface tension as a function of the measured force F and the ring diameter D.

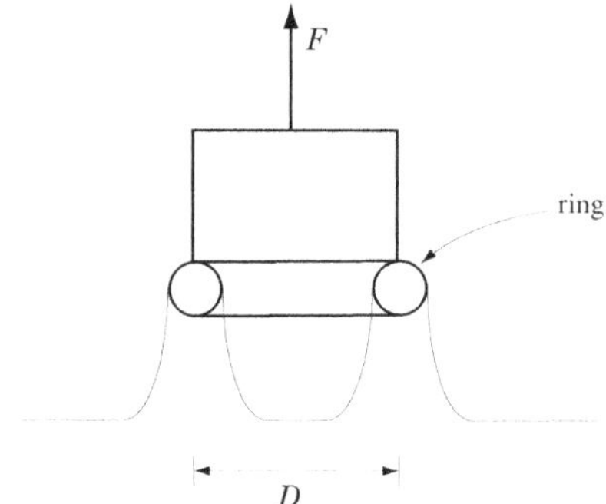

Problem 3.3. Du Noüy balance to measure surface tension.

4

Fluid Statics

Fluid statics is the part of fluid mechanics that deals with fluids when there is no relative motion between the fluid particles. Typically this includes two situations: when the fluid is at rest and when it moves like a rigid solid. This chapter will show how to calculate the pressure field in fluids at rest and how to calculate the interaction forces between the fluid and submerged surfaces.

4.1 The Fundamental Equation of Fluid Statics

Let us take an arbitrary fluid volume V, with surface S. For a fluid at rest, a force balance yields

$$\sum \boldsymbol{F} = \boldsymbol{0}$$

Taking into consideration body and surface forces acting over the volume V,

$$\boldsymbol{F}_v + \boldsymbol{F}_s = \boldsymbol{0} \tag{4.1}$$

that is,

$$\int_V \rho \boldsymbol{f}_m \, \mathrm{d}V + \int_S \boldsymbol{\tau} \boldsymbol{n} \, \mathrm{d}S = \boldsymbol{0} \tag{4.2}$$

In the previous chapter it was shown that a fluid at rest can only withstand normal stresses and, therefore, the stress tensor takes on the form

$$\boldsymbol{\tau} = \begin{bmatrix} -p & 0 & 0 \\ 0 & -p & 0 \\ 0 & 0 & -p \end{bmatrix} = -p\boldsymbol{I} \tag{4.3}$$

where p is the pressure. Substituting,

$$\int_V \rho \boldsymbol{f}_m \, \mathrm{d}V + \int_S -p\boldsymbol{n} \, \mathrm{d}S = \boldsymbol{0} \tag{4.4}$$

and applying the Gauss theorem (see Appendix E) to transform the second integral into a volume integral,

$$\int_V (\rho \boldsymbol{f}_m - \nabla p)\, \mathrm{d}V = \boldsymbol{0} \tag{4.5}$$

Since the above equation holds for any volume, taking the limit as $V \to 0$, the *fundamental equation of fluid statics* is derived,

$$\boxed{\rho \boldsymbol{f}_m - \nabla p = \boldsymbol{0}} \tag{4.6}$$

In Cartesian components,

$$\left\{ \begin{array}{c} \dfrac{\partial p}{\partial x} \\[1ex] \dfrac{\partial p}{\partial y} \\[1ex] \dfrac{\partial p}{\partial z} \end{array} \right\} = \rho \left\{ \begin{array}{c} f_{mx} \\ f_{my} \\ f_{mz} \end{array} \right\} \tag{4.7}$$

Remark 4.1. (a) The net force caused by the pressure depends on the pressure gradient, that is, on spatial variations of pressure. If the pressure is uniform, then it causes no net force over the fluid particle.

(b) The fundamental equation of fluid statics can also be derived from a balance of forces over an infinitesimal fluid cube and applying Taylor series expansions to relate the pressure at opposite sides of the cube.

(c) In rigid solid motions (rotation, translation, or a combination of both), the fundamental equation of fluid statics takes on the form

$$\boxed{\rho \boldsymbol{f}_m - \rho \boldsymbol{a} - \nabla p = \boldsymbol{0}} \tag{4.8}$$

where $\boldsymbol{a}$ is the acceleration of a relative reference system attached to the fluid,

$$\boldsymbol{a} = \underbrace{\boldsymbol{a}_0}_{\text{origin acc.}} + \underbrace{\dot{\boldsymbol{\Omega}} \times \boldsymbol{r}}_{\text{angular acc.}} + \underbrace{\boldsymbol{\Omega} \times (\boldsymbol{\Omega} \times \boldsymbol{r})}_{\text{centripetal acc.}} + \underbrace{2\boldsymbol{\Omega} \times \boldsymbol{v}}_{\text{Coriolis acc.}}$$

$$= \boldsymbol{a}_0 + \dot{\boldsymbol{\Omega}} \times \boldsymbol{r} + \boldsymbol{\Omega} \times (\boldsymbol{\Omega} \times \boldsymbol{r}) \tag{4.9}$$

The last step was given taking into account that the relative velocity $\boldsymbol{v}$ of the fluid particles with respect to the reference system is zero.

4.2 Applications

The applications of fluid statics stem from the fundamental equation (4.6),

$$\nabla p = \rho \boldsymbol{f}_m$$

Next we will cover examples on hydrostatics, manometry and forces on submerged structures.

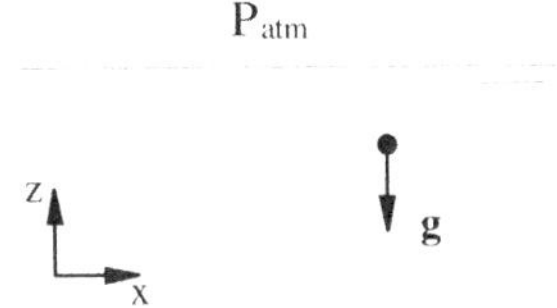

Fig. 4.1. Hydrostatics. Axes.

4.2.1 Hydrostatics

Hydrostatics is the part of fluid statics dedicated to incompressible fluids.

Let us calculate the pressure distribution in a liquid at rest. Take the coordinate axes of Fig. 4.1, where z is the upward vertical axis. The body force due to the acceleration of gravity can be written as

$$\boldsymbol{f}_v = \rho \boldsymbol{f}_m = \rho \boldsymbol{g} = \left\{ \begin{array}{c} 0 \\ 0 \\ -\rho g \end{array} \right\} \tag{4.10}$$

where the density ρ is constant. Substituting into the fundamental equation of fluid statics (4.6)

$$\left\{ \begin{array}{ccc} \dfrac{\partial p}{\partial x} & = & 0 \\[2mm] \dfrac{\partial p}{\partial y} & = & 0 \\[2mm] \dfrac{\partial p}{\partial z} & = & -\rho g \end{array} \right. \tag{4.11}$$

from where the pressure distribution $p(x, y, z)$ can be calculated. The first two equations imply that the pressure is independent of x and y, depending solely on z,

$$p = p(z) \tag{4.12}$$

The third component of the equation can now be written with the total derivative,

$$\frac{\mathrm{d}p}{\mathrm{d}z} = -\rho g \tag{4.13}$$

from where

$$\mathrm{d}p = -\rho g \mathrm{d}z \tag{4.14}$$

Due to the density being constant, the last equation can be integrated between the points 1 and 2 as follows

$$\int_{p_1}^{p_2} \mathrm{d}p = -\rho g \int_{z_1}^{z_2} \mathrm{d}z \tag{4.15}$$

to yield

$$\boxed{p_2 - p_1 = -\rho g(z_2 - z_1)}$$ (4.16)

Remark 4.2. The pressure difference between two points inside a liquid depends only on the height difference between the points.

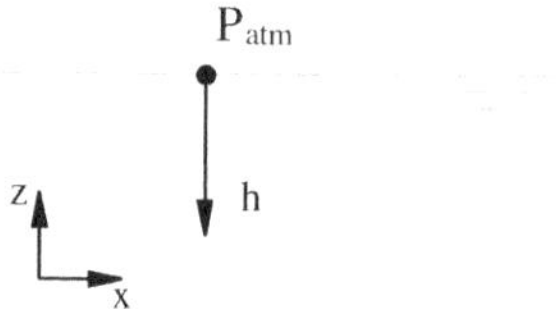

Fig. 4.2. The depth h as a coordinate axis.

The Hydrostatic Pressure as a Function of Depth

Frequently, instead of the vertical axis z, the depth with respect to the free surface h is employed (see Fig. 4.2). In order to obtain the pressure as a function of h, let us introduce the change of variables

$$\mathrm{d}z = -\mathrm{d}h$$ (4.17)

The differential equation (4.14) becomes

$$\mathrm{d}p = \rho g \mathrm{d}h$$ (4.18)

Integrating between a point of depth h and the free surface, where $h = 0$ and $p = p_{\text{atm}}$,

$$\int_{p_{\text{atm}}}^{p} \mathrm{d}p = \rho g \int_{0}^{h} \mathrm{d}h$$

one arrives at

$$\boxed{p = p_{\text{atm}} + \rho g h}$$ (4.19)

Consequences.

(a) The pressure at a point in a liquid depends on the depth of that point with respect to the free surface.
(b) The pressure increases linearly with depth.
(c) The pressure in a liquid does not depend on the shape of the container.

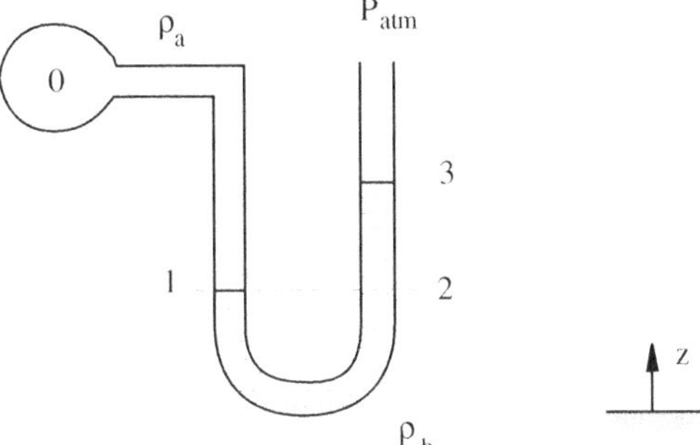

Fig. 4.3. Manometer.

4.2.2 Manometry

A manometer is a device to measure the gage pressure. A typical manometer is depicted in Fig. 4.3.

In order to derive the expression that gives the pressure in a manometer, the hydrostatic equation should be applied within the same fluid, from the point of measurement to the point of reference pressure.

For the manometer of Fig. 4.3, the container pressure at the point O depends on the various liquid levels and equals

$$p_0 = p_{\text{atm}} + \rho_b g(z_3 - z_1) + \rho_a g(z_1 - z_0) \tag{4.20}$$

Example 4.1. Prove equation (4.20).
Solution. Within the same fluid, points at the same height have the same pressure. Thus,

$$p_1 = p_2$$

Next, applying (4.16) within the columns of fluid a and b,

$$p_1 = p_0 + \rho_a g(z_0 - z_1)$$
$$p_2 = p_{\text{atm}} + \rho_b g(z_3 - z_2)$$

Combining the three above equations the desired result is attained.

4.2.3 Fluid Statics of an Isothermal Perfect Gas

When the fluid is a gas, the density is a function of pressure. For a perfect gas (see Appendix H)

$$p = \rho R_{\text{gas}} T \tag{4.21}$$

where R_{gas} is the gas constant, p the absolute pressure, ρ the density and T the absolute temperature. Substituting into the fundamental equation of fluid statics,

$$\mathrm{d}p = -\rho g \mathrm{d}z$$
$$= -\frac{p}{R_{\text{gas}}T} g \mathrm{d}z$$

Separating variables,

$$\frac{\mathrm{d}p}{p} = -\frac{g}{R_{\text{gas}}T}\mathrm{d}z$$

If the temperature variations are neglected,

$$\ln\frac{p_2}{p_1} = -\frac{g}{R_{\text{gas}}T}(z_2 - z_1)$$

$$\boxed{p_2 = p_1\, e^{-\frac{g}{R_{\text{gas}}T}(z_2 - z_1)}} \tag{4.22}$$

4.2.4 Forces over Submerged Surfaces

For a fluid at rest, the surface forces are only due to the pressure. Integration of the stress over the surface yields the force and torque acting on the surface,

$$\boxed{\begin{aligned} \boldsymbol{F} &= \int_S -p\boldsymbol{n}\,\mathrm{d}S \\[2mm] \boldsymbol{M}_O &= \int_S \boldsymbol{r}\times(-p\boldsymbol{n})\,\mathrm{d}S \end{aligned}} \tag{4.23}$$

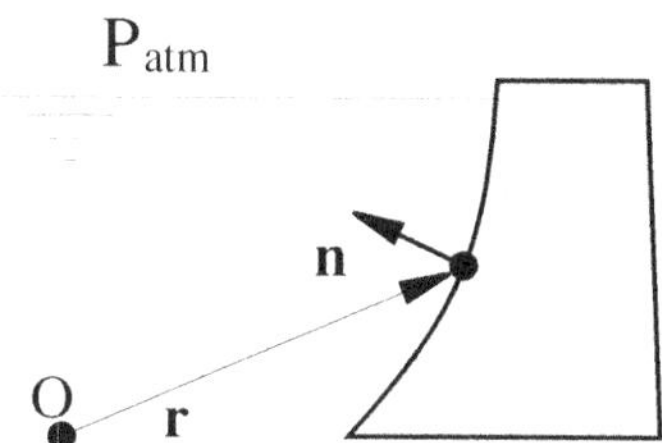

Fig. 4.4. Normal $\boldsymbol{n}$ and position vector $\boldsymbol{r}$ for computing the force and moment with respect to the point O on submerged surfaces.

In the above expressions, $\boldsymbol{n}$ is the exterior normal with respect to the body (see Fig. 4.4), and $\boldsymbol{r}$ is the position vector with respect the point where the moment is calculated.

Center of Pressure

Definition 4.1 (Center of pressure). *The center of pressure is the application point of the resultant force $\boldsymbol{F}$ due to the pressure distribution.*

The center of pressure is calculated imposing that the moment of the resultant force $\boldsymbol{F}$ equals the moment of the pressure distribution with respect to any point,

$$\boldsymbol{r}_{cp} \times \boldsymbol{F} = \int_S \boldsymbol{r} \times (-p\boldsymbol{n}) \, \mathrm{d}S \tag{4.24}$$

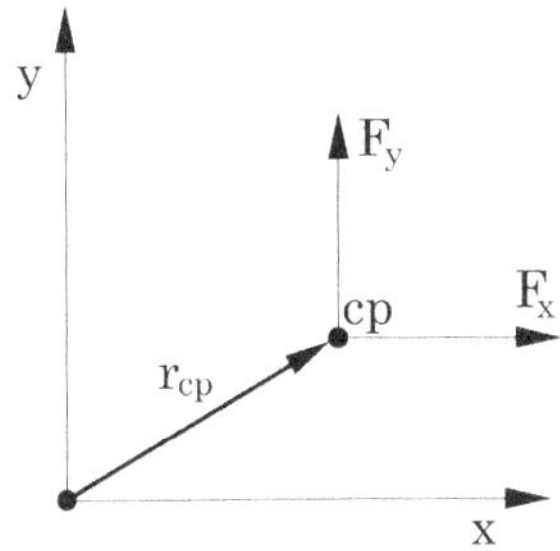

Fig. 4.5. Computation of the moment from the center of pressure.

For two-dimensional flows in the x-y plane, the moment is about the z axis, so the total moment can be calculated from the moments originated by F_x and F_y independently (see Fig. 4.5),

$$M_z = x_{cp} F_y - y_{cp} F_x = \int_S (x\mathrm{d}F_y - y\mathrm{d}F_x) \tag{4.25}$$

Each of the above contributions can be obtained from

$$y_{cp} F_x = \int_S y \, \mathrm{d}F_x$$
$$x_{cp} F_y = \int_S x \, \mathrm{d}F_y \tag{4.26}$$

so the total moment can be calculated as

$$M_z = x_{cp} F_y - y_{cp} F_x \tag{4.27}$$

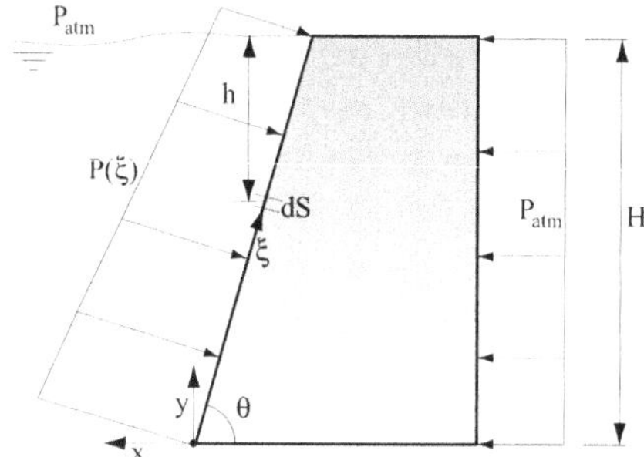

Fig. 4.6. Calculation of the net force on a dam.

Example 4.2 (Net force over a structure). Let us consider the dam of Fig. 4.6. Determine the net force exerted by the hydrostatic pressure and the atmospheric pressure.

Solution. In order to calculate the force due to the hydrostatic pressure on the left side of the dam, we will use the integral

$$\boldsymbol{F}^{\text{left}} = - \int_{S_{\text{left}}} p\boldsymbol{n}\mathrm{d}S$$

On this side,

$$\boldsymbol{n} = \left\{ \begin{array}{c} \sin\theta \\ \cos\theta \end{array} \right\}$$

and

$$p(y) = p_{\text{atm}} + \rho g h = p_{\text{atm}} + \rho g(H - y)$$

To integrate the pressure along the wet surface of the dam, we change variables, $y \to \xi$ and set $\mathrm{d}S = b\mathrm{d}\xi$, with b the width orthogonal to the paper. Next, the pressure distribution needs to be expressed as a function of ξ,

$$p(\xi) = p_{\text{atm}} + \rho g(H - \xi \sin\theta)$$

Substituting,

$$\begin{aligned}
\boldsymbol{F}^{\text{left}} &= - \int_0^{H/\sin\theta} \left[p_{\text{atm}} + \rho g(H - \xi \sin\theta) \right] \left\{ \begin{array}{c} \sin\theta \\ \cos\theta \end{array} \right\} b\mathrm{d}\xi \\
&= -b \left\{ \begin{array}{c} \sin\theta \\ \cos\theta \end{array} \right\} \left[p_{\text{atm}}\xi + \rho g(H\xi - \tfrac{\xi^2}{2}\sin\theta) \right]_0^{H/\sin\theta} \\
&= -b \left\{ \begin{array}{c} \sin\theta \\ \cos\theta \end{array} \right\} \left[p_{\text{atm}}\frac{H}{\sin\theta} + \rho g(\frac{H^2}{2\sin\theta}) \right]
\end{aligned}$$

To get the force due to the atmospheric pressure on the right side of the dam, it will be assumed that the atmospheric pressure is constant and equal to p_{atm}. The force over that side is

$$\boldsymbol{F}^{\text{right}} = -\int_{S_{\text{right}}} p\boldsymbol{n}\mathrm{d}S = -\int_{S_{\text{right}}} p_{\text{atm}} \begin{Bmatrix} -1 \\ 0 \end{Bmatrix} \mathrm{d}S = p_{\text{atm}} H b \begin{Bmatrix} 1 \\ 0 \end{Bmatrix}$$

Similarly, the atmospheric pressure produces on the top and bottom sides a force equal to

$$\boldsymbol{F}^{\text{top}-\text{bottom}} = -\int_{S_{\text{top}-\text{bottom}}} p\boldsymbol{n}\mathrm{d}S = p_{\text{atm}} \frac{Hb}{\tan\theta} \begin{Bmatrix} 0 \\ 1 \end{Bmatrix}$$

The sum of all the contributions gives the *net force*:

$$\boldsymbol{F}^{\text{net}} = -b \begin{Bmatrix} \sin\theta \\ \cos\theta \end{Bmatrix} \left[\rho g \left(\frac{H^2}{2\sin\theta} \right) \right]$$

$$= -\rho g \frac{H^2 b}{2} \begin{Bmatrix} 1 \\ \cotg\,\theta \end{Bmatrix}$$

Corollary 4.1. *When net forces are desired, it is not necessary to consider the atmospheric pressure since, being constant, its action over all the surfaces of the body cancels.*

Example 4.3 (Center of pressure). In the above example, let us compute the center of pressure using (4.26), now without the atmospheric component of the hydrostatic pressure,

$$y_{cp} F_x = \int_S y \,\mathrm{d}F_x$$

$$= -\int_0^{H/\sin\theta} \xi \sin\theta \left[\rho g (H - \xi \sin\theta) \right] \sin\theta b \,\mathrm{d}\xi$$

$$= -b \sin^2\theta \left[\rho g \left(\frac{H\xi^2}{2} - \frac{\xi^3}{3} \sin\theta \right) \right]_0^{H/\sin\theta}$$

$$= -\rho g \frac{H^3 b}{6}$$

Thus

$$y_{cp} = \frac{1}{3} H$$

For the other coordinate of the center of pressure, an analogous process yields

$$x_{cp} F_y = \int_S x \,\mathrm{d}F_y$$

$$= -\int_0^{H/\sin\theta} (-\xi \cos\theta) \left[\rho g (H - \xi \sin\theta) \right] \cos\theta b \,\mathrm{d}\xi$$

$$= b \cos^2\theta \left[\rho g \left(\frac{H\xi^2}{2} - \frac{\xi^3}{3} \sin\theta \right) \right]_0^{H/\sin\theta}$$

$$= \rho g \frac{H^3 b \cotg^2\theta}{6}$$

Thus

$$x_{cp} = -\frac{1}{3} H \cot g\, \theta$$

Finally, the total moment with respect to the coordinates origin is

$$M_z = x_{cp} F_y - y_{cp} F_x = \rho g \frac{H^3 b}{6 \sin^2 \theta}$$

Plane Surfaces

For a plane surface, the resultant force due to the hydrostatic pressure and
the depth of its application point can be calculated as

$$\boxed{\begin{aligned} \boldsymbol{F} &= -p_{cg}\, A\, \boldsymbol{n} \\ h_{cp} &= h_{cg} + \frac{\sin^2 \theta I_{\eta\eta}}{h_{cg} A} \end{aligned}} \tag{4.28}$$

where h_{cg} is the depth of the center of gravity of the surface (centroid) with
respect to the free surface, p_{cg} is the pressure at the center of gravity, A is
the area of the surface, $\boldsymbol{n}$ the normal vector and $I_{\eta\eta}$ the inertia moment of
the surface with respect to a horizontal axis through the center of gravity,

$$I_{\eta\eta} = \int_A \xi^2 \, \mathrm{d}A$$

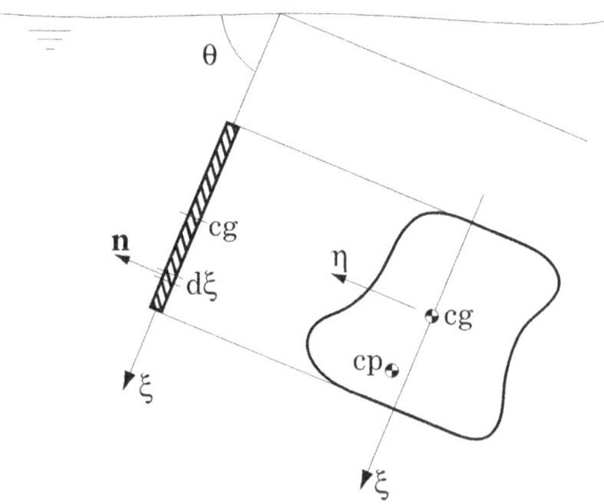

Fig. 4.7. Force and moment of a plane surface.

Table 4.1. Moments of inertia of various surfaces.

Shape	Moment of inertia
	$I_{x'x'} = \frac{1}{12}bh^3$ $I_{xx} = \frac{1}{3}bh^3$
	$I_{xx} = \frac{1}{4}\pi r^4$
	$I_{xx} = \frac{1}{8}\pi r^4$ $I_{yy} = \frac{1}{8}\pi r^4$
	$I_{x'x'} = \frac{1}{36}bh^3$ $I_{xx} = \frac{1}{12}bh^3$
	$I_{xx} = \frac{1}{4}\pi ab^3$ $I_{yy} = \frac{1}{4}\pi a^3 b$

Example 4.4 (Plane surface). Repeat the above example taking into account that the surface of the dam is a plane surface.

Solution. The wetted surface of the dam is a rectangle of area $Hb/\sin\theta$, inclined an angle θ. The center of gravity or geometric center of that rectangle is at the mid-point, which has a depth and a pressure of

$$h_{cg} = \tfrac{1}{2}H$$

$$p_{cg} = \rho g \tfrac{1}{2}H$$

The normal vector, *exterior to the dam*, is

$$n = \left\{ \begin{array}{c} \sin\theta \\ \cos\theta \end{array} \right\}$$

Thus, the net force is given by

$$F = \left\{ \begin{array}{c} F_x \\ F_y \end{array} \right\} = -\rho g \frac{H}{2} \frac{Hb}{\sin\theta} \left\{ \begin{array}{c} \sin\theta \\ \cos\theta \end{array} \right\} = -\rho g \frac{H^2 b}{2} \left\{ \begin{array}{c} 1 \\ \cotg\theta \end{array} \right\}$$

The point of application of the resulting force can be calculated using the moment of inertia of the rectangle, which is

$$I_{\eta\eta} = \frac{1}{12}b(H/\sin\theta)^3$$

Thus,

$$h_{cp} = h_{cg} + \frac{\sin^2\theta\,\frac{1}{12}b(H/\sin\theta)^3}{\frac{1}{2}H(bH/\sin\theta)} = \frac{1}{2}H + \frac{1}{6}H = \frac{2}{3}H$$

Calculation of Forces Using Projected Areas

Sometimes, the calculation of forces and moments due to the pressure distribution can be simplified by using projected areas. This is the case when the horizontal force on a complex surface is searched.

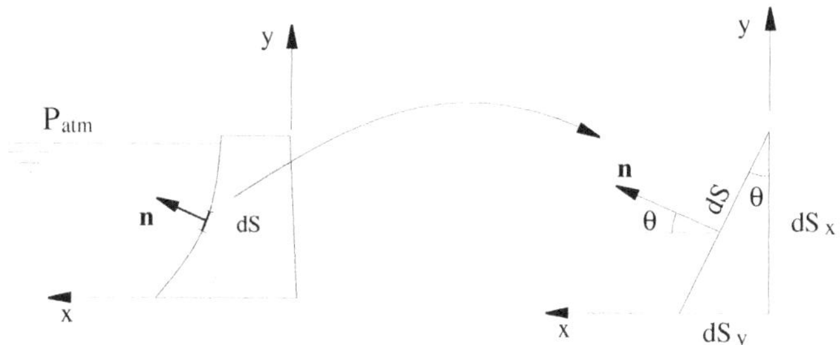

Fig. 4.8. Projection of an infinitesimal area into the coordinate axis.

Let us take a differential area dS on the surface of Fig. 4.8. The projected areas of the differential surface are

$$\begin{aligned}
|n_x|\,\mathrm{d}S &= |\cos\theta|\,\mathrm{d}S = \mathrm{d}S_x > 0 \\
|n_y|\,\mathrm{d}S &= |\sin\theta|\,\mathrm{d}S = \mathrm{d}S_y > 0
\end{aligned} \tag{4.29}$$

where

(a) $\mathrm{d}S_x$ is the projection of $\mathrm{d}S$ orthogonal to the x axis
(b) $\mathrm{d}S_y$ is the projection of $\mathrm{d}S$ orthogonal to the y axis.

In the above expressions, the absolute value is necessary because area is a non-negative magnitude whereas the components of the normal vector are signed.

Then, the force over the surface can be calculated as

$$F_x \;=\; -\int_S p n_x \,\mathrm{d}S$$

$$=\; -\int_S \operatorname{sign}(n_x) p |n_x| \,\mathrm{d}S$$

$$=\; -\int_S \operatorname{sign}(n_x) p \,\mathrm{d}S_x \tag{4.30}$$

and similarly for the y component. Thus,

$$\boxed{\begin{aligned}
F_x &= -\int_S \operatorname{sign}(n_x)\, p \,\mathrm{d}S_x \\[2ex]
F_y &= -\int_S \operatorname{sign}(n_y)\, p \,\mathrm{d}S_y
\end{aligned}} \tag{4.31}$$

Applying the above concepts to the center of pressure, we arrive at

$$\boxed{\begin{aligned}
y_{cp} F_x &= -\int_S \operatorname{sign}(n_x)\, p\,y \,\mathrm{d}S_x \\[2ex]
x_{cp} F_y &= -\int_S \operatorname{sign}(n_y)\, p\,x \,\mathrm{d}S_y
\end{aligned}} \tag{4.32}$$

Corollary 4.2. *The horizontal component of the force F_x does not depend on the shape of the surface.*

Corollary 4.3. *When the pressure is constant, the force that it creates in a given direction equals the pressure times the projected area in that direction.*

Example 4.5 (Projected surface) Determine the horizontal component of the net force F_x exerted by the hydrostatic pressure upon the dam of Fig. 4.8, whose height is H and its width b.

Solution. Using Eqs. (4.31), emanating from the projected areas,

$$\begin{aligned}
F_x &= -\int_S \operatorname{sign}(n_x)\, p \,\mathrm{d}S_x \\[1.5ex]
&= -(+1)\int_S p(y)\, b\,\mathrm{d}y \\[1.5ex]
&= -\int_0^H \rho g(H - y)\, b\,\mathrm{d}y \\[1.5ex]
&= -\rho g\,\frac{H^2 b}{2}
\end{aligned}$$

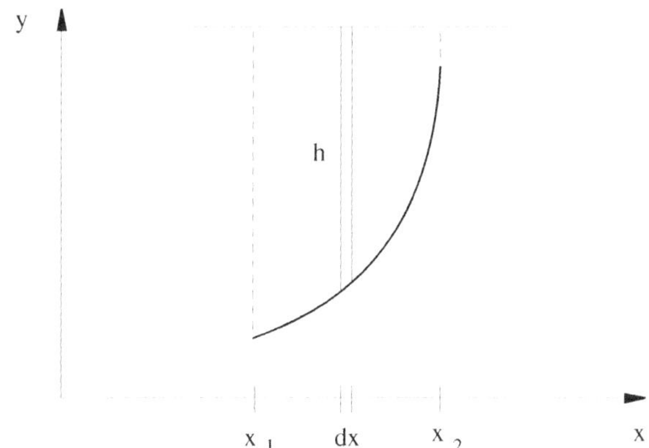

Fig. 4.9. Calculation of the pressure forces as the weight of the fluid between the body and the free surface.

Calculation of Vertical Forces as the Weight of the Volume

Consider the surface of Fig. 4.9. Let us calculate the vertical force exerted over the top side of the surface using the vertically projected area,

$$
\begin{aligned}
F_y &= -\mathrm{sign}(n_y) \int_S \rho g h \, \mathrm{d}S_y \\
&= -\rho g \int_{x_1}^{x_2} h(b\,\mathrm{d}x)
\end{aligned}
\tag{4.33}
$$

The above integral is exactly the volume V of liquid between the surface and the free surface. Thus,

$$
\boxed{F_y = -\rho g V}
\tag{4.34}
$$

The point of application of this force lies at the center of gravity of the volume V.

Example 4.6 (Weight of the fluid volume). In the example of the dam, the vertical force can be determined using the weight of the volume of fluid between the dam and the free surface,

$$
F_y = -\rho g V = -\rho g b \frac{H\,H\,\mathrm{cotg}\,\theta}{2}
$$

Archimedes' Principle

As a consequence of the previous section, the celebrated *Principle of Archimedes* can be deduced: The net buoyant force exerted over a submerged body equals the weight of the displaced fluid.

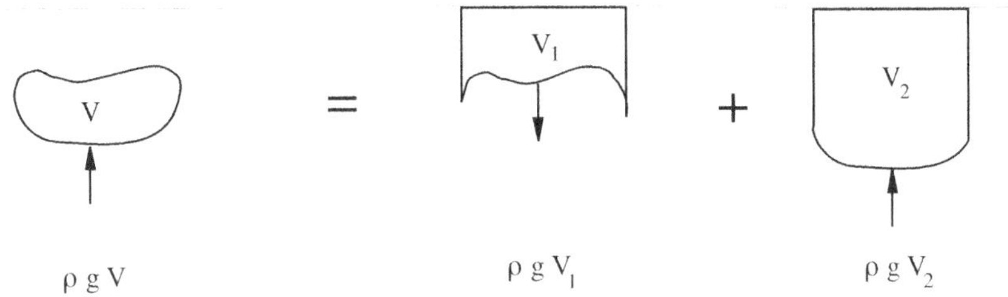

Fig. 4.10. Buoyant force over a submerged body.

The proof of this theorem can be obtained observing Fig. 4.10, where there is an immersed body of volume V. On the top surface the pressure distribution produces a downward force of magnitude $\rho g V_1$, where V_1 is the volume of fluid between the body and the free surface. On the bottom surface there is an upward force of magnitude $\rho g V_2$, where V_2 is the fluid volume between the bottom surface and the free surface. Thus, the net buoyant force is given by

$$\boxed{F = \rho g V_2 - \rho g V_1 = \rho g V} \tag{4.35}$$

Example 4.7 (Archimedes' Principle). Determine the buoyant flow acting upon a sphere of radius $R = 1$ m submerged in water (density $\rho = 1\,000$ kg/m^3).
Solution. Using the Archimedes' Principle,

$$F = \rho g \frac{4}{3}\pi R^3 = 41\,092 \text{ N}$$

Summary

As a summary, there are several ways to calculate forces on submerged structures:

(a) Direct integration
(b) Plane surfaces
(c) Projected areas
(d) Weight of the volume of fluid
(e) Buoyant force: Archimedes' principle

In each particular situation, the simplest method should be chosen.

Problems

4.1   The density ρ of sea water increases as a function of the gage pressure p according to

$$\rho = \rho_0(1 + Kp)$$

The density at sea level ρ_0 (where $p = 0$) is $1\,000$ kg/m^3 and the compressibility coefficient is $K = 5 \times 10^{-10}$ Pa^{-1}. Calculate the pressure at a depth of $5\,000$ m in the following cases:

(a) Assuming that sea water is incompressible, i.e. $K = 0$.
(b) Assuming that the density varies weakly with pressure according to the above expression.

4.2   The inclined gate of the Figure has a hinge at A and is 5 m wide. Determine the net force acting on the inclined surface.

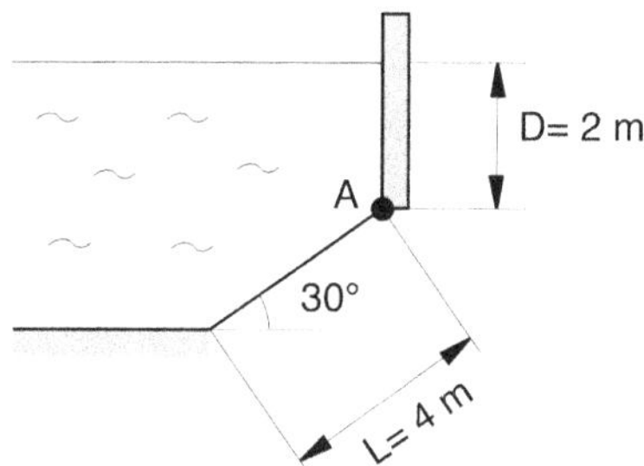

Problem 4.2. Force on a plane gate.

4.3   The gate of a tank articulated at A is shaped as a quarter of a circle. Determine the force per unit width T necessary to keep the gate in the depicted position.

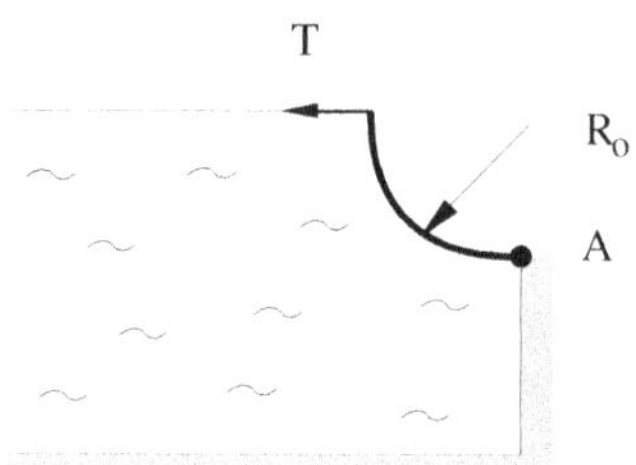

Problem 4.3. Force acting on a circular gate.

4.4 The gate ABC of the Figure has the shape of an L and it is articulated at B. Its width perpendicular to the paper is 2 m. When the water level is sufficiently high, the gate opens at A, allowing the liquid to flow. Calculate the water level h at which this happens.

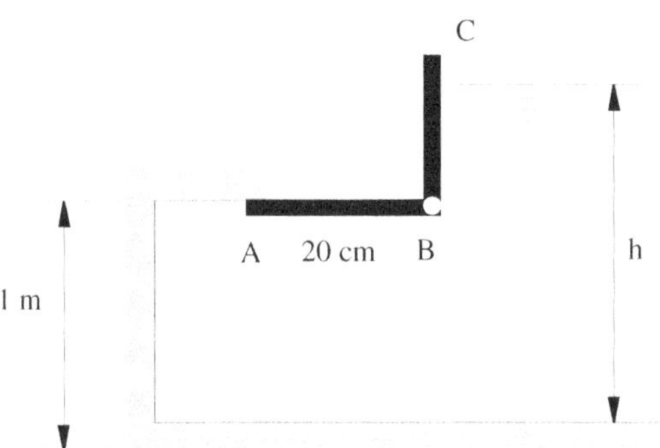

Problem 4.4. Automatic gate.

4.5 The square gate of the Figure has a side of 1 m and is articulated in B. It automatically opens if the water level h is high enough. Calculate the minimum level for opening. Does the result depend on the liquid density?

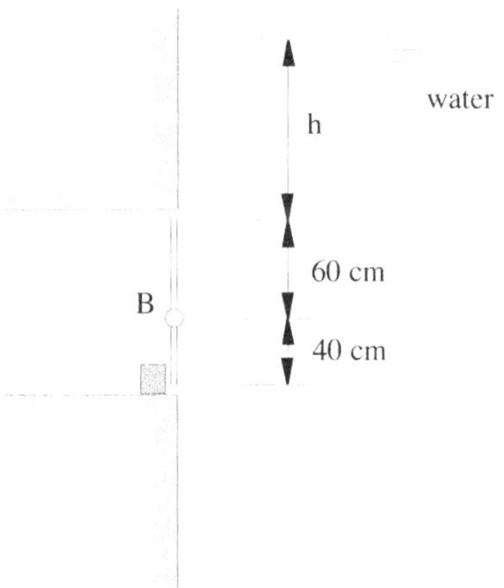

Problem 4.5. Plane automatic gate.

4.6   Before launching a weather balloon, it is filled with 100 kg of helium at atmospheric pressure and temperature (1.033×10^5 Pa and 288.15 K, respectively).

(a) Calculate the maximum load (balloon plus scientific load) that can be carried by the above amount of helium.
(b) The balloon rises with the load of item (a) up to an altitude of 11 km, where the atmospheric conditions are 2.263×10^4 Pa and 216.65 K. If the helium pressure and temperature are equal to the atmospheric conditions, calculate the net force acting upon the balloon and scientific load at this altitude.

Gas constants: $R_{\mathrm{He}} = 2\,077$ J/(kg K), $R_{\mathrm{air}} = 287$ J/(kg K)

4.7   Prove the result (4.28) for plane surfaces. Use that

$$\xi_{cp} = \frac{\sin \theta \, I_{\eta\eta}}{h_{cg} A}$$

4.8   Relate the reading of the inclined manometer of the Figure with the pressure at O.

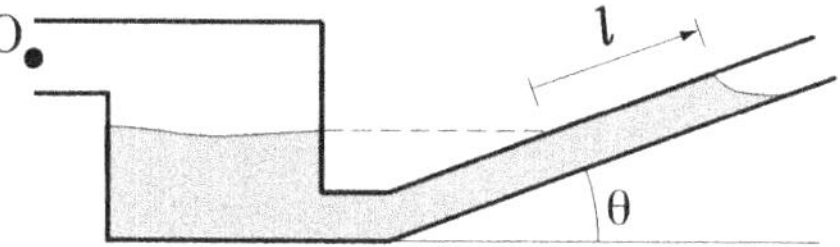

Problem 4.8. Inclined manometer.

4.9   The pressure loss $\Delta p = p_1 - p_2$ in the duct of the Figure is due to friction losses. Assuming fully developed flow, calculate the pressure loss as a function of the fluid densities and h.

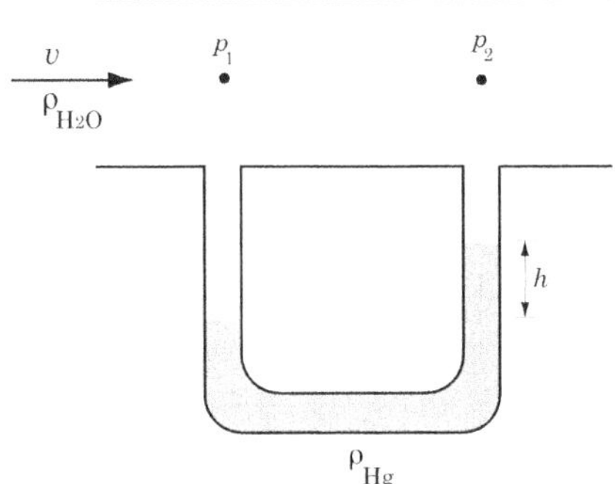

Problem 4.9. Pressure loss in a horizontal duct.

Part II

Conservation Principles

5

Transport Theorems

The application of classical mechanics principles to fluid dynamics is not a straightforward task. Since fluids are extremely deformable media, Lagrangian formulations find a hard time tracking all the fluid particles. Therefore, it is very convenient to rewrite the equations of mechanics and thermodynamics in a more amenable way for the solution of transport problems. This can be achieved with the help of the transport theorems, presented in this chapter.

5.1 Fluid Volume and Control Volume

The fundamental laws of mechanics, such as the second Newton's law, have been initially developed for particle systems or pieces of matter. In transport phenomena, we find ourselves with pieces of matter that deform in complex ways, continuously changing shape and size. Therefore, given an initial set of fluid particles, it is a difficult task to follow their motion and evolution.

Fig. 5.1. Evolution of a fluid volume $V_f(t)$.

Definition 5.1 (Fluid volume). *A fluid volume is a volume that always contains the same particles of fluid. It is denoted by $V_f(t)$. In continuum* *mechanics, it is also called a material volume, because it is formed by and follows the same piece of matter. As a consequence, a fluid volume moves with the fluid velocity, $\boldsymbol{v}$.*

However, to solve practical problems, it would be very convenient that the equations could be applied to any volume, arbitrarily chosen. Portions of this volume could be fixed or in motion; matter could flow across its boundaries or its boundaries could follow the fluid; and so on. This gives rise to the concept of control volume.

Definition 5.2 (Control volume). *A control volume is an arbitrary volume selected to analyze a transport problem. It is denoted by $V_c(t)$ and it moves at the velocity $\boldsymbol{v}^c$.*

In principle, the control volumes used in this text will contain only fluid, but this is not the only approach. A control volume can contain any parts of the analyzed system, such as walls and other devices.

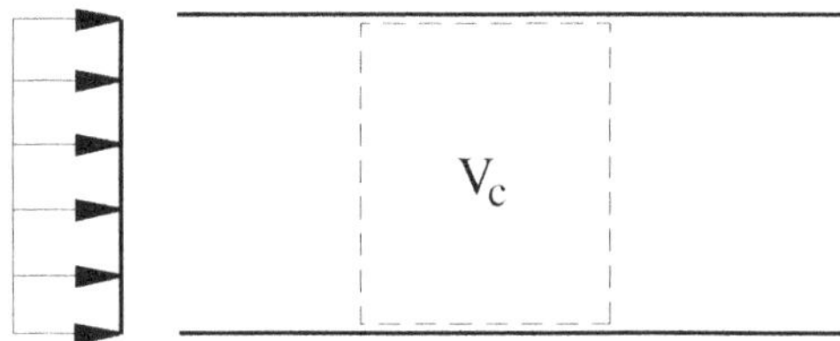

Fig. 5.2. Fixed control volume $V_c(t)$ to analyze, for example, a pipe.

The relationship between the equations written for a fluid volume and those written for a control volume is based on the Reynolds transport theorems, which are derived next.

5.2 Transport Theorems

The transport integral equations are made of integrals of the form

$$\frac{\mathrm{d}}{\mathrm{d}t} \int_{V(t)} f(\boldsymbol{x}, t) \, \mathrm{d}V \tag{5.1}$$

One can notice that

(a) The domain integral $V(t)$ depends on time t.
(b) The integrand $f(\cdot, t)$ also depends on time t.

Therefore, when taking the derivative of (5.1) with respect to time, two contributions will emanate. One way to perform this calculation is using the Leibnitz differentiation rule for integrals.

Theorem 5.1. *Let $V(t)$ be an arbitrary volume with boundary $S(t)$, which moves at the velocity $\boldsymbol{w}$. The exterior normal to the boundary is denoted by $\boldsymbol{n}$. Then,*

$$\frac{\mathrm{d}}{\mathrm{d}t} \int_{V(t)} \phi(\boldsymbol{x}, t)\, \mathrm{d}V = \int_{V(t)} \frac{\partial \phi(\boldsymbol{x}, t)}{\partial t}\, \mathrm{d}V + \int_{S(t)} \phi(\boldsymbol{x}, t)\, \boldsymbol{w} \cdot \boldsymbol{n}\, \mathrm{d}S \qquad (5.2)$$

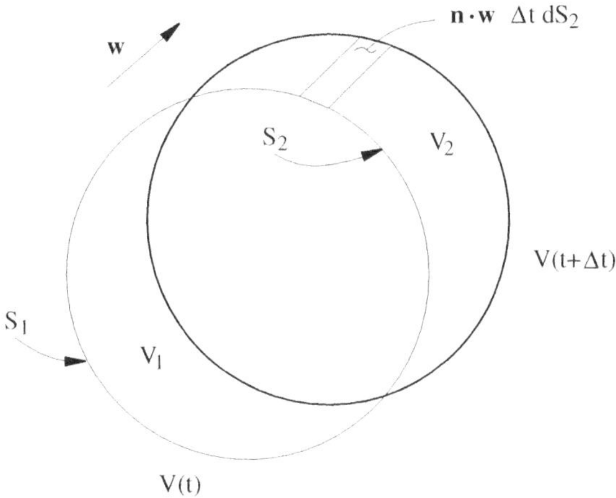

Fig. 5.3. Moving domain of integration $V(t)$ with surface $S(t)$.

Proof. Let $V(t)$ be the domain at time t, shown in Fig. 5.3, whose surface moves at a speed $\boldsymbol{w}$. At some later time $t + \Delta t$, the domain will occupy the space denoted by $V(t + \Delta t)$. This volume at $t + \Delta t$ can be decomposed as

$$V(t + \Delta t) = V(t) + V_2(\Delta t) - V_1(\Delta t) \qquad (5.3)$$

The intersection of the surfaces of $V(t)$ and $V(t + \Delta t)$, decomposes $S(t)$ into two surfaces, $S_1(t)$ y $S_2(t)$, as shown in the Figure.

Applying the definition of derivative,

$$\frac{\mathrm{d}}{\mathrm{d}t} \int_{V(t)} \phi(\boldsymbol{x}, t)\, \mathrm{d}V = \lim_{\Delta t \to 0} \frac{1}{\Delta t} \left[\int_{V(t+\Delta t)} \phi(\boldsymbol{x}, t + \Delta t)\, \mathrm{d}V - \int_{V(t)} \phi(\boldsymbol{x}, t)\, \mathrm{d}V \right]$$
$$(5.4)$$

Next, the decomposition (5.3) is substituted in the integral on $V(t + \Delta t)$

$$
\frac{\mathrm{d}}{\mathrm{d}t} \int_{V(t)} \phi(\boldsymbol{x},t)\ \mathrm{d}V \;=\; \lim_{\Delta t \to 0} \frac{1}{\Delta t}\Bigg[\int_{V(t)} \phi(\boldsymbol{x},t+\Delta t)\ \mathrm{d}V
$$

$$
+ \int_{V_2(\Delta t)} \phi(\boldsymbol{x},t+\Delta t)\ \mathrm{d}V - \int_{V_1(\Delta t)} \phi(\boldsymbol{x},t+\Delta t)\ \mathrm{d}V
$$

$$
- \int_{V(t)} \phi(\boldsymbol{x},t)\ \mathrm{d}V \Bigg] \tag{5.5}
$$

By definition of derivative, the combination of the first and last integrals is precisely

$$
\int_{V(t)} \frac{\partial \phi(\boldsymbol{x},t)}{\partial t}\ \mathrm{d}V \tag{5.6}
$$

Now let us turn our attention to the second integral on the right-hand side of (5.5). The volume differential in $V_2(\Delta t)$ can be written as

$$
\mathrm{d}V = \text{base} \times \text{height} = \mathrm{d}S\ (\boldsymbol{w}\Delta t)\cdot\boldsymbol{n} \tag{5.7}
$$

where both the volume differential and $(\boldsymbol{w}\Delta t)\cdot\boldsymbol{n}$ are positive. Thus,

$$
\lim_{\Delta t \to 0} \frac{1}{\Delta t} \int_{V_2(\Delta t)} \phi(\boldsymbol{x},t+\Delta t)\ \mathrm{d}V
$$

$$
= \lim_{\Delta t \to 0} \frac{1}{\Delta t} \int_{S_2(t)} \phi(\boldsymbol{x},t+\Delta t)\ (\boldsymbol{w}\Delta t\cdot\boldsymbol{n})\ \mathrm{d}S
$$

$$
= \int_{S_2(t)} \phi(\boldsymbol{x},t)\ \boldsymbol{w}\cdot\boldsymbol{n}\ \mathrm{d}S \tag{5.8}
$$

Analogously for the integral on V_1, where in this case the volume differential is positive but the product $(\boldsymbol{w}\Delta t)\cdot\boldsymbol{n}$, negative. Therefore,

$$
\mathrm{d}V = \text{base} \times \text{height} = -\mathrm{d}S\ (\boldsymbol{w}\Delta t)\cdot\boldsymbol{n} \tag{5.9}
$$

Thus,

$$
\lim_{\Delta t \to 0} -\frac{1}{\Delta t} \int_{V_1(\Delta t)} \phi(\boldsymbol{x},t+\Delta t)\ \mathrm{d}V
$$

$$
= \lim_{\Delta t \to 0} \frac{1}{\Delta t} \int_{S_1(t)} \phi(\boldsymbol{x},t+\Delta t)\ (\boldsymbol{w}\Delta t\cdot\boldsymbol{n})\ \mathrm{d}S
$$

$$
= \int_{S_1(t)} \phi(\boldsymbol{x},t)\ \boldsymbol{w}\cdot\boldsymbol{n}\ \mathrm{d}S \tag{5.10}
$$

Gathering all the contributions, the desired result is attained.

5.2.1 First Transport Theorem

This first theorem consists of applying the Reynolds transport theorem to a fluid volume. Thus, choosing $V(t) = V_f(t)$, $S(t) = S_f(t)$ and $\boldsymbol{w} = \boldsymbol{v}$,

$$
\boxed{\ \frac{\mathrm{d}}{\mathrm{d}t} \int_{V_f(t)} \phi(\boldsymbol{x},t)\ \mathrm{d}V = \int_{V_f(t)} \frac{\partial \phi(\boldsymbol{x},t)}{\partial t}\ \mathrm{d}V + \int_{S_f(t)} \phi(\boldsymbol{x},t)\ \boldsymbol{v}\cdot\boldsymbol{n}\ \mathrm{d}S\ } \tag{5.11}
$$

5.2.2 Second Transport Theorem

It consists of applying the Reynolds transport theorem to a control volume. Thus, selecting $V(t) = V_c(t)$, $S(t) = S_c(t)$ and $\boldsymbol{w} = \boldsymbol{v}^c$,

$$\frac{\mathrm{d}}{\mathrm{d}t} \int_{V_c(t)} \phi(\boldsymbol{x}, t)\, \mathrm{d}V = \int_{V_c(t)} \frac{\partial \phi(\boldsymbol{x}, t)}{\partial t}\, \mathrm{d}V + \int_{S_c(t)} \phi(\boldsymbol{x}, t)\, \boldsymbol{v}^c \cdot \boldsymbol{n}\, \mathrm{d}S \qquad (5.12)$$

5.2.3 Third Transport Theorem

The third transport theorem relates the rate of variation of integrals on a fluid and a control volume. Let $V_c(t)$ be an arbitrary control volume. Next, choose a fluid volume $V_f(t)$ such that it coincides with the control volume at time t:

$$V_f(t) = V_c(t) \qquad (5.13)$$

Then, the following integrals are equal,

$$\int_{V_f(t)} \frac{\partial \phi(\boldsymbol{x}, t)}{\partial t}\, \mathrm{d}V = \int_{V_c(t)} \frac{\partial \phi(\boldsymbol{x}, t)}{\partial t}\, \mathrm{d}V \qquad (5.14)$$

Finally, subtracting the first two transport theorems yields,

$$\frac{\mathrm{d}}{\mathrm{d}t} \int_{V_f(t)} \phi(\boldsymbol{x}, t)\, \mathrm{d}V = \frac{\mathrm{d}}{\mathrm{d}t} \int_{V_c(t)} \phi(\boldsymbol{x}, t)\, \mathrm{d}V + \int_{S_c(t)} \phi(\boldsymbol{x}, t)\, (\boldsymbol{v} - \boldsymbol{v}^c) \cdot \boldsymbol{n}\, \mathrm{d}S$$

$$(5.15)$$

In this manner, we have related integrals over fluid volumes to integrals over control volumes. The first term on the right-hand side reflects the variation in time of the property $\phi(\boldsymbol{x}, t)$ inside the control volume, $V_c(t)$. The second term on the right-hand side represents the contribution due to the flux across the control volume surface $S_c(t)$.

6

Integral Conservation Principles

The equations that govern the dynamics of fluids stem from the application of the basic laws of mechanics and thermodynamics to moving deformable media. In this chapter, from these basic laws, the integral conservation equations of fluid dynamics will be derived. These include the equation of mass conservation, the equations of momentum and angular momentum, and the equation of energy conservation. Special attention is given to various forms of energy equations. Also, the equation of mass conservation will be extended to multicomponent reactive systems to derive the conservation equations of chemical species. Finally, a useful equation for immiscible liquids will be presented.

The equations derived in this chapter apply for multicomponent systems with the same body force for all chemical species. For the transport equations where the body force depends on the species, see Appendix I.

6.1 Mass Conservation

The principle of mass conservation establishes that the mass of a fluid volume (a volume that always contains the same fluid particles) is constant. Therefore, for a fluid volume,

$$\text{Mass}(V_f) = \text{constant} \tag{6.1}$$

and

$$\frac{\mathrm{d}}{\mathrm{d}t}\text{Mass}(V_f) = 0 \tag{6.2}$$

The mass of a fluid volume can be written as

$$\text{Mass}(V_f) = \int_{V_f(t)} \rho \, \mathrm{d}V \tag{6.3}$$

and substituting into the above equation,

$$\boxed{\frac{\mathrm{d}}{\mathrm{d}t} \int_{V_f(t)} \rho \, \mathrm{d}V = 0} \tag{6.4}$$

which is the equation of mass conservation for a fluid volume.

Applying the third transport theorem with $\phi = \rho$ (see Chapter 5), the equation is transformed for a control volume,

$$\boxed{\frac{\mathrm{d}}{\mathrm{d}t} \int_{V_c(t)} \rho \, \mathrm{d}V + \int_{S_c(t)} \rho \left[(\boldsymbol{v} - \boldsymbol{v}^c) \cdot \boldsymbol{n} \right] \mathrm{d}S = 0} \tag{6.5}$$

Rearranging the convective term into the right-hand side, the physical interpretation of the above equation is that the rate of change of the mass in the control volume, $\frac{\mathrm{d}}{\mathrm{d}t} M(V_c)$, equals the ingoing minus the outgoing mass flow rate (kg/s) across the surface, $\dot{M}_{\mathrm{in}} - \dot{M}_{\mathrm{ou}}$. In other words, this equation represents the global balance of mass per unit time

$$\frac{\mathrm{d}}{\mathrm{d}t} M(V_c) = \dot{M}_{\mathrm{in}} - \dot{M}_{\mathrm{out}} \tag{6.6}$$

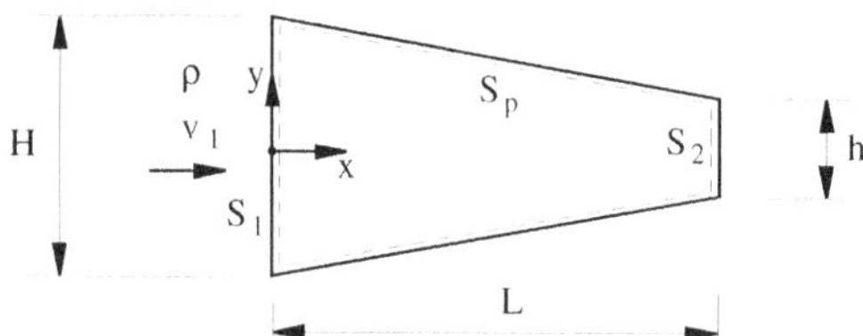

Fig. 6.1. One-dimensional flow in a converging nozzle.

Example 6.1 (One-dimensional steady flow through a converging nozzle). Along the square cross-section nozzle of Fig. 6.1, the side of length H at the inlet decreases to h at the outlet. If the liquid density ρ and the inlet velocity v_1 are known, determine the exit uniform fluid velocity v_2.

Solution. Let us apply the mass conservation equation. The first step consists of choosing a control volume for analysis, which in this example contains the fluid within the nozzle between sections 1 and 2. Since this is a fixed control volume, $\boldsymbol{v}^c = \boldsymbol{0}$. The surface of the control volume is split into the surfaces S_1 (inlet), S_2 (outlet), S_p (walls). Then,

$$\underbrace{\frac{\mathrm{d}}{\mathrm{d}t} \int_{V_c(t)} \rho \, \mathrm{d}V}_{\text{steady flow}} + \int_{S_1+S_2+S_p} \rho \left[(\boldsymbol{v} - \boldsymbol{v}^c) \cdot \boldsymbol{n} \right] \mathrm{d}S = 0$$

The integral along S_p vanishes because there is no mass flow through solid walls. At the walls, either the velocity is zero (viscous flow) or the velocity is parallel to the wall (ideal flow). If the flow is one-dimensional, the velocity at each cross section of the nozzle is uniform and the integrals are easily calculated,

$$-\rho v_1 H^2 + \rho v_2 h^2 = 0$$

from where the exit velocity is

$$v_2 = \frac{H^2}{h^2} v_1$$

Example 6.2 (Velocity distribution along a nozzle). In the previous example, find the one-dimensional velocity distribution along the nozzle $v(x)$.

Solution. Again, the mass conservation equation is applied to the control volume between the inlet section S_1 at $x = 0$ and an arbitrary cross section $S(x)$ at x. Repeating the above process,

$$\underbrace{\frac{\mathrm{d}}{\mathrm{d}t} \int_{V_c(t)} \rho \, \mathrm{d}V}_{\text{steady flow}} + \int_{S_1 + S(x) + S_p} \rho \left[(\boldsymbol{v} - \boldsymbol{v}^c) \cdot \boldsymbol{n} \right] \mathrm{d}S = 0$$

and integrating,

$$-\rho v_1 H^2 + \rho v(x) S(x) = 0$$

Assuming a linear variation of the side of the nozzle with x,

$$S(x) = \left(H - \frac{H - h}{L} x \right)^2$$

and substituting, the desired velocity is obtained,

$$v(x) = \frac{1}{\left[1 - \left(1 - \frac{h}{H} \right) \frac{x}{L} \right]^2} \, v_1$$

Example 6.3 (Pumped fluid by a piston pump). The piston inside a cylinder of diameter D moves at a speed V. Assuming that the fluid is incompressible, determine the relationship between the velocity of the piston V and the outflow velocity of the fluid v_s at the exhaust valve of diameter d.

Solution. The solution is found by application of the equation of mass conservation. In this case, the control volume contains all the fluid inside the cylinder, which is a moving control volume, $\boldsymbol{v}^c \neq \mathbf{0}$. Furthermore, in this problem the flow is unsteady. The surface of the control volume is divided

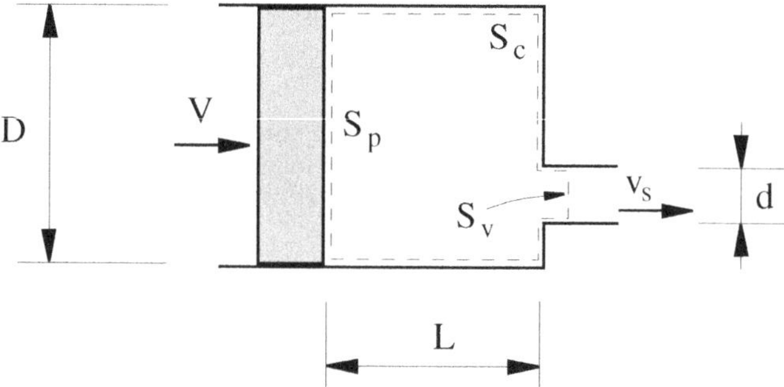

Fig. 6.2. Pumped flow by a piston pump.

into S_p (moving wall of the piston), S_v (outflow section at the exhaust valve), S_c (fixed walls of the cylinder). Then,

$$\frac{\mathrm{d}}{\mathrm{d}t} \int_{V_c(t)} \rho \, \mathrm{d}V + \int_{S_p+S_v+S_c} \rho \left[(\boldsymbol{v} - \boldsymbol{v}^c) \cdot \boldsymbol{n} \right] \mathrm{d}S = 0$$

where the integral on S_c cancels because there is no mass flux across the solid wall. Operating,

$$\frac{\mathrm{d}}{\mathrm{d}t} \int_{V_c(t)} \rho \, \mathrm{d}V = \rho \frac{\mathrm{d}}{\mathrm{d}t} V_c(t) = \rho \frac{\pi D^2}{4} \frac{\mathrm{d}L}{\mathrm{d}t} = -\rho \frac{\pi D^2}{4} V$$

$$\int_{S_p} \rho \left[(\boldsymbol{v} - \boldsymbol{v}^c) \cdot \boldsymbol{n} \right] \mathrm{d}S = 0$$
$$\underbrace{\qquad\qquad}_{\boldsymbol{v}=\boldsymbol{v}^c}$$

$$\int_{S_v} \rho \left[(\boldsymbol{v} - \boldsymbol{v}^c) \cdot \boldsymbol{n} \right] \mathrm{d}S = \rho v_s \frac{\pi d^2}{4}$$

Note that, even though S_p is in motion, there is no mass flow across the piston because the fluid velocity equals the piston velocity. The last integral has been evaluated assuming a uniform velocity profile at the exit S_v. Gathering all the contributions,

$$v_s = \frac{D^2}{d^2} V$$

6.2 Momentum Equation

This equation stems from the fundamental equation of mechanics, the second law of Newton,

$$\boldsymbol{P} = \sum \boldsymbol{F}_{\text{ext}} \tag{6.7}$$

where P is the linear momentum and $\sum F_{\text{ext}}$, the external forces acting over the particles of the system.

Recall that for a system of particles $P = \sum_i m_i v_i$ and, therefore, for a fluid volume, the sum extended to an infinite number of particles of mass $dm = \rho dV$ is equivalent to an integral,

$$P = \int_{V_f(t)} \rho v \, dV \tag{6.8}$$

In Chapter 3 it was shown that, neglecting the surface tension, the forces acting on the fluid particle are body and surface forces,

$$\sum F_{\text{ext}} = \int_{S_f(t)} f_s \, dS + \int_{V_f(t)} \rho f_m \, dV \tag{6.9}$$

$$= \int_{S_f(t)} \tau n \, dS + \int_{V_f(t)} \rho f_m \, dV \tag{6.10}$$

Thus, the momentum integral equation of a fluid volume can be written as

$$\boxed{\frac{d}{dt} \int_{V_f(t)} \rho v \, dV = \int_{S_f(t)} \tau n \, dS + \int_{V_f(t)} \rho f_m \, dV} \tag{6.11}$$

Applying the third transport theorem with $\phi = \rho v$, the following equation for a control volume is derived

$$\boxed{\frac{d}{dt} \int_{V_c(t)} \rho v \, dV + \int_{S_c(t)} \rho v \left[(v - v^c) \cdot n \right] dS = \int_{S_c(t)} \tau n \, dS + \int_{V_c(t)} \rho f_m \, dV}$$
$$\tag{6.12}$$

The physical meaning of this equation is that the rate of change of the momentum inside the $V_c(t)$ plus the outgoing flux of momentum across the surface of the control volume $\dot{P}_{\text{out}} - \dot{P}_{\text{in}}$ equals the action of the surface and body forces over the control volume. Rearranging,

$$\frac{d}{dt} P(V_c) = \dot{P}_{\text{in}} - \dot{P}_{\text{out}} + \sum F_{\text{ext}} \tag{6.13}$$

Remark 6.1. The integral

$$\int_{V_c(t)} \rho f_m \, dV$$

for $f_m = g$ represents the weight of the fluid inside the control volume $V_c(t)$.

6.2.1 Decomposition of the Stress Tensor

In fluid statics, typically when the fluid velocities are zero, it was seen that the stress tensor is an isotropic tensor, given by the static pressure (see Section 3.3.2),

$$\boldsymbol{\tau} = \begin{pmatrix} -p & 0 & 0 \\ 0 & -p & 0 \\ 0 & 0 & -p \end{pmatrix} = -p\boldsymbol{I} \tag{6.14}$$

In the presence of a velocity field, the stress tensor presents an additional term due to the viscous stresses of the velocity field, $\boldsymbol{\tau}'$, so that the total stress tensor is the sum of both contributions,

$$\boxed{\boldsymbol{\tau} = -p\boldsymbol{I} + \boldsymbol{\tau}'} \tag{6.15}$$

The tensor $\boldsymbol{\tau}'$ is called the *viscous stress tensor*, and it is responsible for the friction that real fluids undergo. *Real* fluids always present friction. In contrast, fluids without friction are called *ideal* fluids. The property that measures this quality of the fluid is called *viscosity* μ and will be described in more detail in Chapter 7 together with the constitutive equation for $\boldsymbol{\tau}'$.

Therefore, the surface forces on a real fluid can be expressed as

$$\int_{S_c(t)} \boldsymbol{f}_s \, \mathrm{d}S = \int_{S_c(t)} \boldsymbol{\tau}\boldsymbol{n} \, \mathrm{d}S = \int_{S_c(t)} (-p\boldsymbol{n} + \boldsymbol{\tau}'\boldsymbol{n}) \, \mathrm{d}S \tag{6.16}$$

Example 6.4 (Force to hold a symmetric bifurcation). A flow in a circular cross-section pipe is divided into two streams, as Fig. 6.3 depicts. The flow is steady with density $\rho = 1\,000$ kg/m^3. Neglecting the force due to gravity, determine the necessary force to hold the bifurcation.

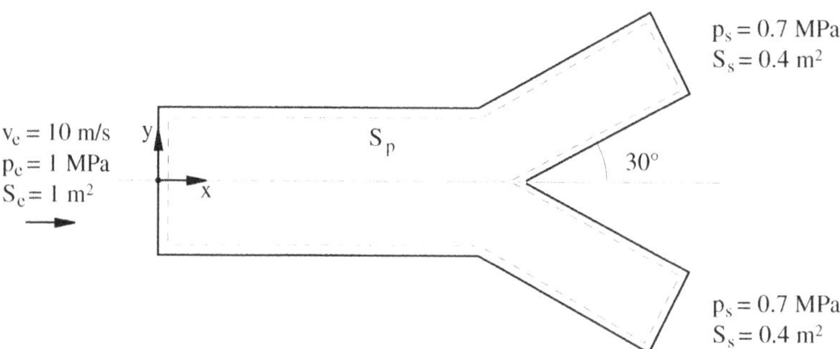

Fig. 6.3. Symmetric flow in a bifurcation.

Solution. Let us select the control volume between the inlet section, S_e, and the outlet sections $2S_s$ of the bifurcation. The remaining surfaces of the control volume correspond to solid surfaces (walls), S_p. To determine the exit velocity v_s, which will be assumed aligned with the duct, the mass conservation equation is applied assuming uniform fluid properties at inlets and outlets, in particular, velocities and pressures. The result is

$$v_s = \frac{v_e S_e}{2 S_s} = 12.5 \text{ m/s}$$

Due to the symmetry of the flow with respect to the horizontal axis, and having ignored the weight of the fluid, the vertical forces cancel. Thus, it is only necessary to apply the x component of the momentum equation, which is extracted as follows,

$$\underbrace{\frac{\mathrm{d}}{\mathrm{d}t} \int_{V_c(t)} \rho v_x \, \mathrm{d}V}_{\text{stationary}} + \int_{S_e + 2S_s + S_p} \rho v_x \left[(\boldsymbol{v} - \boldsymbol{v}^c) \cdot \boldsymbol{n} \right] \mathrm{d}S = \int_{S_c(t)} \boldsymbol{\tau n} \Big|_x \, \mathrm{d}S$$

$$+ \underbrace{\int_{V_c(t)} \rho f_{mx} \, \mathrm{d}V}_{\text{no } x \text{ component}}$$

Thus, the left-hand side equals

$$\int_{S_e + 2S_s} \rho v_x \left(\boldsymbol{v} \cdot \boldsymbol{n} \right) \mathrm{d}S = \rho v_e (-v_e S_e) + 2 \, \rho \, (v_s \cos 30°)(v_s S_s)$$

In the surface forces term, the interaction force from the solid walls to the fluid is gathered in $\boldsymbol{F}_{\text{wall}\to\text{fluid}}$. Due to the third Newton's law (action-reaction), this force has the same magnitude, but opposite sign, as the force exerted by the flow to the solid wall. Thus,

$$\boxed{\int_{S_p} \left(-p\boldsymbol{n} + \boldsymbol{\tau}'\boldsymbol{n} \right) \mathrm{d}S = \boldsymbol{F}_{\text{wall}\to\text{fluid}} = -\boldsymbol{F}_{\text{fluid}\to\text{wall}}}$$

Taking everything into account, the right-hand side can be written as

$$\int_{S_c(t)} \boldsymbol{\tau n} \Big|_x \, \mathrm{d}S = \int_{S_e + 2S_s} -p n_x \, \mathrm{d}S + \underbrace{\int_{S_e + 2S_s} \boldsymbol{\tau}'\boldsymbol{n} \Big|_x \, \mathrm{d}S}_{\tau' \approx 0 \text{ inlet, outlet}}$$

$$+ \underbrace{\int_{S_p} \left(-p n_x + \boldsymbol{\tau}'\boldsymbol{n} \Big|_x \right) \mathrm{d}S}_{F_x = F_{\text{wall}\to\text{fluid},x}}$$

$$= -p_e(-1)S_e - 2 p_s \cos 30° \, S_s + F_x$$

Substituting, F_x is found to be

$$F_x = \rho(2 v_s^2 S_s \cos 30° - v_e^2 S_e) + 2 p_s \cos 30° \, S_s - p_e S_e$$

$$= -507 \text{ kN}$$

6.3 Angular Momentum Equation

The angular momentum equation for a system of particles establishes that

$$\frac{\mathrm{d}}{\mathrm{d}t} \boldsymbol{H} = \sum \boldsymbol{M}_{\mathrm{ext}}$$

where $\boldsymbol{H}$ represents the angular momentum and $\sum \boldsymbol{M}_{\mathrm{ext}}$, the moment of the external forces acting over the particles. For a system of particles, the angular momentum is $\boldsymbol{H} = \sum_i m_i (\boldsymbol{r}_i \times \boldsymbol{v}_i)$. Thus, for a fluid volume, integrating over all the fluid particles of mass $\rho \mathrm{d}V$,

$$\boldsymbol{H} = \int_{V_f(t)} \boldsymbol{r} \times \rho \boldsymbol{v} \, \mathrm{d}V \tag{6.17}$$

The moment over the fluid particle is due to surface and body forces, respectively,

$$\mathrm{d}\boldsymbol{M}_s = \boldsymbol{r} \times \mathrm{d}\boldsymbol{F}_s = \boldsymbol{r} \times \boldsymbol{f}_s \, \mathrm{d}S \tag{6.18}$$

$$\mathrm{d}\boldsymbol{M}_v = \boldsymbol{r} \times \mathrm{d}\boldsymbol{F}_v = \boldsymbol{r} \times \rho \boldsymbol{f}_m \, \mathrm{d}V \tag{6.19}$$

The total moment is the sum of both contributions,

$$\sum \boldsymbol{M}_{\mathrm{ext}} = \boldsymbol{M}_s + \boldsymbol{M}_v \tag{6.20}$$

$$= \int_{S_f(t)} \boldsymbol{r} \times \boldsymbol{f}_s \, \mathrm{d}S + \int_{V_f(t)} \boldsymbol{r} \times \rho \boldsymbol{f}_m \, \mathrm{d}V \tag{6.21}$$

Substituting yields,

$$\boxed{\frac{\mathrm{d}}{\mathrm{d}t} \int_{V_f(t)} \boldsymbol{r} \times \rho \boldsymbol{v} \, \mathrm{d}V = \int_{S_f(t)} \boldsymbol{r} \times (\boldsymbol{\tau} \cdot \boldsymbol{n}) \, \mathrm{d}S + \int_{V_f(t)} \boldsymbol{r} \times \rho \boldsymbol{f}_m \, \mathrm{d}V} \tag{6.22}$$

And for a control volume,

$$\boxed{\begin{aligned} \frac{\mathrm{d}}{\mathrm{d}t} \int_{V_c(t)} \boldsymbol{r} \times \rho \boldsymbol{v} \, \mathrm{d}V &+ \int_{S_c(t)} \boldsymbol{r} \times \rho \boldsymbol{v} \left[(\boldsymbol{v} - \boldsymbol{v}^c) \cdot \boldsymbol{n} \right] \mathrm{d}S \\ &= \int_{S_c(t)} \boldsymbol{r} \times (\boldsymbol{\tau} \cdot \boldsymbol{n}) \, \mathrm{d}S + \int_{V_c(t)} \boldsymbol{r} \times \rho \boldsymbol{f}_m \, \mathrm{d}V \end{aligned}}$$

$$\tag{6.23}$$

The physical meaning of this equation is that the rate of variation of the angular momentum per unit time inside $V_c(t)$ equals the net inflow angular momentum across the surfaces of the control volume (product of the volumetric flux times the angular momentum per unit volume) plus the moment of the external forces. Mathematically,

$$\frac{\mathrm{d}}{\mathrm{d}t} \boldsymbol{H}(V_c) = \dot{\boldsymbol{H}}_{\mathrm{in}} - \dot{\boldsymbol{H}}_{\mathrm{out}} + \sum \boldsymbol{M}_{\mathrm{ext}} \tag{6.24}$$

Example 6.5 (Force and moment to hold an inclined jet). Let us calculate the force $\boldsymbol{F}$ and moment $\boldsymbol{M}$ to hold the nozzle of Fig. 6.4, which has a constant cross-section S.

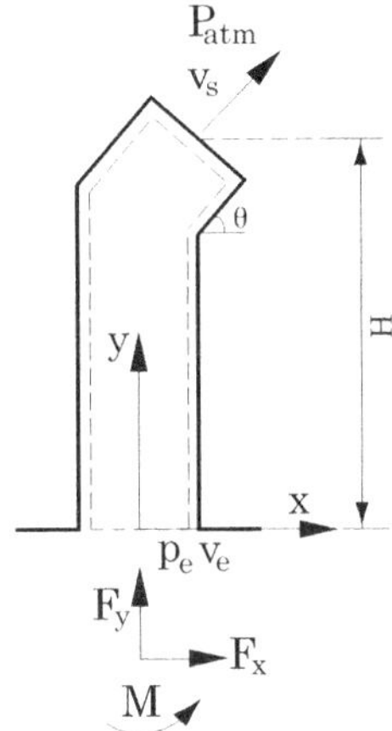

Fig. 6.4. Reactions at the base of an inclined nozzle.

Solution. We will assume a stationary, incompressible flow with uniform properties at the inlet and outlet. The viscous stresses will be neglected at the inlet and outlet. The gravity forces are also ignored, which means that the moment due to the weight of the fluid inside the nozzle is neglegible compared to the pressure moment or change of the fluid angular momentum. Finally, gage pressures will be used, i.e. $p_{atm} = 0$.

Then, let us take the fixed control volume of the Figure, which includes the fluid inside the nozzle. Its boundaries are the inlet S_e, outlet S_s and solid walls S_p.

The mass conservation equation implies

$$v_e = v_s = \frac{Q}{S}$$

The momentum equation can be used to determine the force to hold the nozzle,

$$\underbrace{\frac{\mathrm{d}}{\mathrm{d}t} \int_{V_c(t)} \rho \boldsymbol{v} \, \mathrm{d}V}_{\text{steady}} + \int_{S_e + S_s + S_p} \rho \boldsymbol{v} \left[(\boldsymbol{v} - \boldsymbol{v}^c) \cdot \boldsymbol{n} \right] \mathrm{d}S = \int_{S_c(t)} \boldsymbol{\tau} \boldsymbol{n} \, \mathrm{d}S + \underbrace{\int_{V_c(t)} \rho \boldsymbol{f}_m \, \mathrm{d}V}_{\text{ignored}}$$

Taking into account that $\boldsymbol{v}_e = (0, v_e, 0)$ and $\boldsymbol{v}_s = v_s(\cos\theta, \sin\theta, 0)$, the momentum flux is

$$\int_{S_e+S_s} \rho \boldsymbol{v}\,[\boldsymbol{v}\cdot\boldsymbol{n}]\,\mathrm{d}S = \rho \left\{ \begin{array}{c} 0 \\ v_e \\ 0 \end{array} \right\} (-Q) + \rho v_s \left\{ \begin{array}{c} \cos\theta \\ \sin\theta \\ 0 \end{array} \right\} (+Q)$$

The surface forces are calculated as follows:

$$\int_{S_e+S_s+S_p} \boldsymbol{\tau n}\,\mathrm{d}S \;=\; \underbrace{\int_{S_p} \boldsymbol{\tau n}\,\mathrm{d}S}_{\boldsymbol{F}_{\text{wall}\to\text{fluid}}}$$

$$+ \int_{S_e} -p\boldsymbol{n}\,\mathrm{d}S + \underbrace{\int_{S_s} -p\boldsymbol{n}\,\mathrm{d}S}_{p_{\text{atm}}=0}$$

$$+ \underbrace{\int_{S_e+S_s} \boldsymbol{\tau}'\boldsymbol{n}\;\mathrm{d}S}_{\boldsymbol{\tau}'\approx 0\,\text{inlet, outlet}}$$

$$= \boldsymbol{F}_{\text{wall}\to\text{fluid}} - p_e \left\{ \begin{array}{c} 0 \\ -1 \\ 0 \end{array} \right\} S_e$$

Finally, the reactions at the base are

$$\boldsymbol{F}_{\text{wall}\to\text{fluid}} = \left\{ \begin{array}{c} F_x \\ F_y \end{array} \right\} = \rho Q \left\{ \begin{array}{c} v_s \cos\theta \\ -v_s(1-\sin\theta) \\ 0 \end{array} \right\} - \left\{ \begin{array}{c} 0 \\ p_e S \\ 0 \end{array} \right\}$$

Thus, the nozzle tends to move to the left and upwards, and the forces to hold it point to the right and downwards.

Next, to calculate the moment at the base, we apply the angular momentum equation,

$$\underbrace{\frac{\mathrm{d}}{\mathrm{d}t}\int_{V_c(t)} \boldsymbol{r}\times\rho\boldsymbol{v}\,\mathrm{d}V}_{\text{steady}} + \int_{S_e+S_s+S_p} \boldsymbol{r}\times\rho\boldsymbol{v}\,[(\boldsymbol{v}-\boldsymbol{v}^c)\cdot\boldsymbol{n}]\,\mathrm{d}S \;=\; \int_{S_c(t)} \boldsymbol{r}\times\boldsymbol{\tau n}\,\mathrm{d}S$$

$$+ \underbrace{\int_{V_c(t)} \boldsymbol{r}\times\rho\boldsymbol{f}_m\,\mathrm{d}V}_{\text{ignored}}$$

Taking moments with respect to the origin of coordinates, at S_e, $\boldsymbol{r}_e \approx \boldsymbol{0}$, and at S_s we will take $\boldsymbol{r}_s \approx (0,H,0)$. As $\boldsymbol{v}_s = v_s(\cos\theta,\sin\theta,0)$,

$$\int_{S_e+S_s} \boldsymbol{r}\times\rho\boldsymbol{v}\,[\boldsymbol{v}\cdot\boldsymbol{n}]\,\mathrm{d}S = \boldsymbol{0} + \left\{ \begin{array}{c} 0 \\ H \\ 0 \end{array} \right\} \times \rho v_s \left\{ \begin{array}{c} \cos\theta \\ \sin\theta \\ 0 \end{array} \right\} (+Q) = \left\{ \begin{array}{c} 0 \\ 0 \\ -\rho H v_s \cos\theta\, Q \end{array} \right\}$$

Similarly, for the surface forces,

$$\int_{S_c+S_s+S_p} \boldsymbol{r} \times \boldsymbol{\tau} \boldsymbol{n} \, \mathrm{d}S = \underbrace{\int_{S_p} \boldsymbol{r} \times \boldsymbol{\tau} \boldsymbol{n} \, \mathrm{d}S}_{\boldsymbol{M}_{\mathrm{wall}\to\mathrm{fluid}}}$$

$$+ \underbrace{\int_{S_c} \boldsymbol{r} \times -p\boldsymbol{n} \, \mathrm{d}S}_{r_c=0} + \underbrace{\int_{S_s} \boldsymbol{r} \times -p\boldsymbol{n} \, \mathrm{d}S}_{p_{\mathrm{atm}}=0}$$

$$+ \underbrace{\int_{S_c+S_s} \boldsymbol{r} \times \boldsymbol{\tau}'\boldsymbol{n} \, \mathrm{d}S}_{\boldsymbol{\tau}'\approx 0 \,\mathrm{inlet,\,outlet}}$$

$$= \boldsymbol{M}_{\mathrm{wall}\to\mathrm{fluid}}$$

Gathering all the contributions, the resulting moment (about the z axis) is

$$\boldsymbol{M}_{\mathrm{wall}\to\mathrm{fluid}} = \left\{ \begin{array}{c} 0 \\ 0 \\ -\rho H v_s \cos\theta Q \end{array} \right\}$$

The negative sign implies that the moment is towards the right.

6.4 Total Energy Conservation

The total energy equation is based on the first principle of thermodynamics: the change of total energy E in a system equals the work done over the system W_{ext} plus the added heat Q_{in}:

$$\Delta E_{\mathrm{system}} = W_{\mathrm{ext}} + Q_{\mathrm{in}} \tag{6.25}$$

Rewriting the balance per unit time and taking the limit

$$\lim_{\Delta t \to 0} \frac{\Delta E_{\mathrm{system}}}{\Delta t} = \frac{\mathrm{d}E}{\mathrm{d}t} \tag{6.26}$$

which gives the equation in its rate form, with dimension of power,

$$\frac{\mathrm{d}E}{\mathrm{d}t} = \dot{W}_{\mathrm{ext}} + \dot{Q}_{\mathrm{in}} \tag{6.27}$$

For a simple compressible substance, the possible forms of energy are the internal and kinetic energy,

$$E = \int_{V_f(t)} \rho e_{\mathrm{tot}} \, \mathrm{d}V = \int_{V_f(t)} \rho(e + \frac{1}{2}v^2) \, \mathrm{d}V$$

where e_{tot} represents the total energy per unit mass, e [J/kg] the specific internal energy (see Appendix H) and v the velocity modulus, that is, $v^2 = \boldsymbol{v} \cdot \boldsymbol{v}$.

On the other side, the power of the exterior forces is,

$$\dot{W}_{\text{ext}} = \int_{S_f(t)} (\boldsymbol{\tau}\boldsymbol{n}) \cdot \boldsymbol{v} \, \mathrm{d}S + \int_{V_f(t)} \rho \boldsymbol{f}_m \cdot \boldsymbol{v} \, \mathrm{d}V$$

The heat transmitted to the system has two contributions. One is due to thermal diffusion across the surface, which can be calculated from the *heat flux* vector, $\boldsymbol{q}$ (see Chapter 2). The second contribution is caused by the heat received per unit volume, $\dot{q}_v$. Thus,

$$\dot{Q}_{\text{in}} = \int_{S_f(t)} -\boldsymbol{q} \cdot \boldsymbol{n} \, \mathrm{d}S + \int_{V_f(t)} \dot{q}_v \, \mathrm{d}V$$

The minus sign on the heat flux term is due to the fact that the contribution represents heat entering the system, towards the fluid, therefore, in the direction $-\boldsymbol{n}$. The corresponding SI units are $[\dot{q}_v] = \mathrm{W/m^3}$, $[\boldsymbol{q}] = \mathrm{W/m^2}$.

Thus, for a fluid volume, we have

$$\boxed{\begin{aligned}
\frac{\mathrm{d}}{\mathrm{d}t} \int_{V_f(t)} \rho \left(e + \frac{1}{2}v^2 \right) \mathrm{d}V &= \int_{S_f(t)} (-p\boldsymbol{n} + \boldsymbol{\tau}'\boldsymbol{n}) \cdot \boldsymbol{v} \, \mathrm{d}S \\
&+ \int_{V_f(t)} \rho \boldsymbol{f}_m \cdot \boldsymbol{v} \, \mathrm{d}V \\
&- \int_{S_f(t)} \boldsymbol{q} \cdot \boldsymbol{n} \, \mathrm{d}S + \int_{V_f(t)} \dot{q}_v \, \mathrm{d}V
\end{aligned}} \tag{6.28}$$

Applying the third transport theorem,

$$\boxed{\begin{aligned}
\frac{\mathrm{d}}{\mathrm{d}t} \int_{V_c(t)} \rho \left(e + \frac{1}{2}v^2 \right) \mathrm{d}V &+ \int_{S_c(t)} \rho \left(e + \frac{1}{2}v^2 \right) [(\boldsymbol{v} - \boldsymbol{v}^c) \cdot \boldsymbol{n}] \, \mathrm{d}S = \\
&= \int_{S_c(t)} (-p\boldsymbol{n} + \boldsymbol{\tau}'\boldsymbol{n}) \cdot \boldsymbol{v} \, \mathrm{d}S \\
&+ \int_{V_c(t)} \rho \boldsymbol{f}_m \cdot \boldsymbol{v} \, \mathrm{d}V \\
&- \int_{S_c(t)} \boldsymbol{q} \cdot \boldsymbol{n} \, \mathrm{d}S + \int_{V_c(t)} \dot{q}_v \, \mathrm{d}V
\end{aligned}}$$

$$\tag{6.29}$$

From the physical standpoint, the above equation represents the following energy balance

$$\frac{\mathrm{d}}{\mathrm{d}t} E(V_c) = \dot{E}_{\text{in}} - \dot{E}_{\text{out}} + \dot{W}_{\text{ext}} + \dot{Q}_{\text{in}} \tag{6.30}$$

That is, the rate of increase of the total energy in the control volume is due to the incoming energy flux, minus the outgoing energy flux, plus the work done on the system per unit time plus the added heat per unit time.

6.4.1 Body Force Stemming from a Potential

When the body force $\boldsymbol{f}_m$ stems from a potential, it can be written as the gradient of a scalar function U, called the *potential*,

$$\boldsymbol{f}_m = -\nabla U \tag{6.31}$$

The most common case is that of the gravitational force, which for an upward z axis,

$$U = gz \tag{6.32}$$

It can be checked that the gradient of the potential gives

$$-\nabla U = \left\{ \begin{array}{c} 0 \\ 0 \\ -g \end{array} \right\} \tag{6.33}$$

which is indeed the force per unit mass $\boldsymbol{f}_m = \boldsymbol{g}$.

To transform the body force integral into a potential energy term, we will employ the continuity equation (see Chapter 8)

$$\frac{\partial \rho}{\partial t} + \nabla \cdot (\rho \boldsymbol{v}) = 0$$

Starting from the integral of the body force power,

$$\int_{V_c(t)} \rho \; \boldsymbol{f}_m \cdot \boldsymbol{v} \; \mathrm{d}V = \int_{V_c(t)} -\rho \nabla U \cdot \boldsymbol{v} \; \mathrm{d}V \quad \text{(potential definition)}$$

$$= \int_{V_c(t)} \left(-\nabla \cdot (\rho U \boldsymbol{v}) + U \nabla \cdot (\rho \boldsymbol{v}) \right) \; \mathrm{d}V \quad \text{(product differential)}$$

$$= \int_{V_c(t)} \left(-\nabla \cdot (\rho U \boldsymbol{v}) - U \frac{\partial \rho}{\partial t} \right) \; \mathrm{d}V \quad \text{(continuity)}$$

$$= \int_{V_c(t)} \left(-\nabla \cdot (\rho U \boldsymbol{v}) - \frac{\partial \rho U}{\partial t} \right) \; \mathrm{d}V \quad (U \text{ not a function of } t)$$

$$= -\int_{V_c(t)} \frac{\partial \rho U}{\partial t} \; \mathrm{d}V - \int_{S_c(t)} \rho U \boldsymbol{v} \cdot \boldsymbol{n} \; \mathrm{d}S \quad \text{(Gauss theorem)}$$

$$- \int_{S_c(t)} \rho U \boldsymbol{v}^c \cdot \boldsymbol{n} \; \mathrm{d}S + \int_{S_c(t)} \rho U \boldsymbol{v}^c \cdot \boldsymbol{n} \; \mathrm{d}S \quad \text{(add \& substract)}$$

$$= -\frac{\mathrm{d}}{\mathrm{d}t} \int_{V_c(t)} \rho U \; \mathrm{d}V - \int_{S_c(t)} \rho U (\boldsymbol{v} - \boldsymbol{v}^c) \cdot \boldsymbol{n} \; \mathrm{d}S \quad \text{(transp. th.)}$$

Inserting this term into the total energy equation, if the potential is steady,

$$\boxed{\begin{aligned} &\frac{\mathrm{d}}{\mathrm{d}t} \int_{V_c(t)} \rho \left(e + \tfrac{1}{2} v^2 + U \right) \; \mathrm{d}V \\ &\quad + \int_{S_c(t)} \rho \left(e + \frac{1}{2} v^2 + U \right) \left[(\boldsymbol{v} - \boldsymbol{v}^c) \cdot \boldsymbol{n} \right] \; \mathrm{d}S = \\ &\quad = \int_{S_c(t)} (-p\boldsymbol{n} + \boldsymbol{\tau}' \cdot \boldsymbol{n}) \cdot \boldsymbol{v} \; \mathrm{d}S - \int_{S_c(t)} \boldsymbol{q} \cdot \boldsymbol{n} \; \mathrm{d}S + \int_{V_c(t)} \dot{q}_v \; \mathrm{d}V \end{aligned}} \tag{6.34}$$

Remark 6.2. In the energy equations, the body force must be accounted only once: either as a potential energy term U or as the power of the body forces.

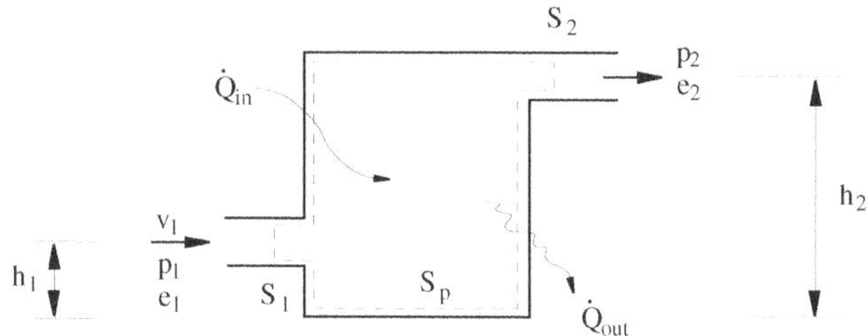

Fig. 6.5. Principle of total energy conservation in a burner.

Example 6.6 (Heat added to a burner). A device adds $\dot{Q}_{in}$ heat per unit time to the burner of Fig. 6.5, so the specific internal energy of the fluid is increased from e_1 to e_2 at the exit . Determine the heat losses per unit time $\dot{Q}_{out}$ through the walls.

Solution. Let us start applying the mass conservation equation with the usual hypothesis: steady, incompressible flow, with uniform inlet and outlet properties.

The fixed control volume (depicted in Fig. 6.5) includes the fluid inside the burner. Applying the mass conservation equation, the exit velocity results in

$$v_2 = \frac{v_1 S_1}{S_2} = \frac{Q}{S_2}$$

where Q is the incoming volumetric flux. Next, the total energy conservation principle states

$$\underbrace{\frac{\mathrm{d}}{\mathrm{d}t} \int_{V_c(t)} \rho \left(e + \frac{1}{2} v^2 + U \right) \, \mathrm{d}V}_{\text{steady}}$$

$$+ \int_{S_e + S_s + S_p} \rho \left(e + \frac{1}{2} v^2 + U \right) \left[(\boldsymbol{v} - \boldsymbol{v}^c) \cdot \boldsymbol{n} \right] \mathrm{d}S$$

$$= \int_{S_e(t)} (-p\boldsymbol{n} + \boldsymbol{\tau}' \cdot \boldsymbol{n}) \cdot \boldsymbol{v} \, \mathrm{d}S \underbrace{- \int_{S_e(t)} \boldsymbol{q} \cdot \boldsymbol{n} \, \mathrm{d}S}_{\dot{Q}_{in} - \dot{Q}_{out}} + \underbrace{\int_{V_c(t)} \dot{q}_v \, \mathrm{d}V}_{\text{absent, considered in } \boldsymbol{q}}$$

The energy flux and the stress integrals can be calculated as

$$\int_{S_c+S_s} \rho \left(e + \frac{1}{2}v^2 + U \right) (\boldsymbol{v} \cdot \boldsymbol{n})\, \mathrm{d}S = \quad -\rho(e_1 + \tfrac{1}{2}v_1^2 + gh_1)v_1 S_1$$

$$+\rho(e_2 + \tfrac{1}{2}v_2^2 + gh_2)v_2 S_2$$

$$\int_{S_1+S_2+S_p} -p\boldsymbol{n} \cdot \boldsymbol{v}\, \mathrm{d}S = p_1 v_1 S_1 - p_2 v_2 S_2 + \underbrace{\int_{S_p} -p\boldsymbol{n} \cdot \boldsymbol{v}\, \mathrm{d}S}_{\boldsymbol{v}=\boldsymbol{0}}$$

$$\int_{S_1+S_2+S_p} (\boldsymbol{\tau}' \cdot \boldsymbol{n}) \cdot \boldsymbol{v}\, \mathrm{d}S = \underbrace{\int_{S_1+S_2} (\boldsymbol{\tau}' \cdot \boldsymbol{n}) \cdot \boldsymbol{v}\, \mathrm{d}S}_{\boldsymbol{\tau}' \approx 0 \; \text{inlet, outlet}} + \underbrace{\int_{S_p} (\boldsymbol{\tau}' \cdot \boldsymbol{n}) \cdot \boldsymbol{v}\, \mathrm{d}S}_{\boldsymbol{v}=\boldsymbol{0} \; \text{on walls if } \mu \neq 0}$$

The integral $\int_{S_p} (\boldsymbol{\tau}' \cdot \boldsymbol{n}) \cdot \boldsymbol{v}\, \mathrm{d}S$ cancels also under the condition of ideal flow, in which case $\mu = 0$ and, consequently, $\boldsymbol{\tau}' = 0$. In an ideal flow, the velocity is parallel to the wall and $-\int_{S_p} p\boldsymbol{n} \cdot \boldsymbol{v}\, \mathrm{d}S = 0$. Gathering everything, one gets $\dot{Q}_{\mathrm{out}}$ as

$$\dot{Q}_{\mathrm{out}} = \dot{Q}_{\mathrm{in}} + \rho(e_1 + \tfrac{1}{2}v_1^2 + gh_1)v_1 S_1 - \rho(e_2 + \tfrac{1}{2}v_2^2 + gh_2)v_2 S_2$$

$$+ p_1 v_1 S_1 - p_2 v_2 S_2$$

$$= \dot{Q}_{\mathrm{in}} + \rho Q \left[(e_1 + \tfrac{1}{2}v_1^2 + gh_1 + \tfrac{p_1}{\rho}) - (e_2 + \tfrac{1}{2}v_2^2 + gh_2 + \tfrac{p_2}{\rho}) \right]$$

That is, the heat lost per unit time is the difference between the total energy that the fluid should have at the exit minus the total energy that the fluid actually has. Note that the pressure enters into the energy balance because due to the fluid motion it exchanges power.

6.5 Other Energy Equations

6.5.1 Mechanical Energy Equation

The mechanical energy equation can be obtained from the differential form of the momentum equation (see Chapter 8),

$$\frac{\partial \rho \boldsymbol{v}}{\partial t} + \nabla \cdot (\rho \boldsymbol{v}\boldsymbol{v}) = -\nabla p + \nabla \cdot \boldsymbol{\tau}' + \rho \boldsymbol{f}_m \tag{6.35}$$

This equation is multiplied by the scalar product of the vector velocity $\boldsymbol{v}$

$$\boldsymbol{v} \cdot \left(\frac{\partial \rho \boldsymbol{v}}{\partial t} + \nabla \cdot (\rho \boldsymbol{v}\boldsymbol{v}) = -\nabla p + \nabla \cdot \boldsymbol{\tau}' + \rho \boldsymbol{f}_m \right) \tag{6.36}$$

Operating yields

$$\frac{\partial \rho \frac{1}{2} v^2}{\partial t} + \nabla \cdot (\rho \boldsymbol{v} \frac{1}{2} v^2) = -\boldsymbol{v} \cdot \nabla p + \boldsymbol{v} \cdot \nabla \cdot \boldsymbol{\tau}' + \rho \boldsymbol{v} \cdot \boldsymbol{f}_m \tag{6.37}$$

The first term on the right-hand side can be rewritten as

$$\boldsymbol{v} \cdot \nabla p = \nabla \cdot (p\boldsymbol{v}) - p\nabla \cdot \boldsymbol{v} \tag{6.38}$$

which equals the *power of the pressure forces* $\nabla \cdot (p\boldsymbol{v})$ minus the *expansion power*,

$$p\nabla \cdot \boldsymbol{v} = p\, v_{i,i} \tag{6.39}$$

The reader is reminded that the divergence of the velocity represents the change of volume per unit time per unit volume of a fluid particle,

$$\text{div}\ \boldsymbol{v} = \nabla \cdot \boldsymbol{v} = v_{i,i} = \frac{1}{\text{Vol}} \frac{\text{dVol}}{\text{d}t}$$

The second term on the right-hand side can be rewritten as

$$\begin{aligned}
\boldsymbol{v} \cdot \nabla \cdot \boldsymbol{\tau}' &= v_i\, \tau'_{ij,j} = (v_i \tau'_{ij})_{,j} - v_{i,j} \tau'_{ij} \\
&= \nabla \cdot (\boldsymbol{\tau}' \boldsymbol{v}) - \phi_v
\end{aligned} \tag{6.40}$$

The function ϕ_v is called the *viscous dissipation function*,

$$\boxed{\phi_v = \nabla \boldsymbol{v} : \boldsymbol{\tau}' = v_{i,j}\, \tau'_{ij} \geq 0} \tag{6.41}$$

which is always *positive*. Remember that the Einstein summation convention is used and repeated indices are added up. In Cartesian coordinates, the viscous dissipation function can be expanded as

$$\begin{aligned}
\phi_v =\ & v_{1,1}\, \tau'_{11} + v_{1,2}\, \tau'_{12} + v_{1,3}\, \tau'_{13} \\
+\ & v_{2,1}\, \tau'_{21} + v_{2,2}\, \tau'_{22} + v_{2,3}\, \tau'_{23} \\
+\ & v_{3,1}\, \tau'_{31} + v_{3,2}\, \tau'_{32} + v_{3,3}\, \tau'_{33}
\end{aligned} \tag{6.42}$$

The viscous dissipation represents the energy loss per unit time and volume due to the viscosity, that is, to friction. Its units are power per unit volume, W/m^3.

Integrating in a fluid volume and applying the transport theorems,

$$\boxed{\begin{aligned}
\frac{\text{d}}{\text{d}t} \int_{V_c(t)} \rho \tfrac{1}{2} v^2\ \text{d}V &+ \int_{S_c(t)} \rho \frac{1}{2} v^2\, [(\boldsymbol{v} - \boldsymbol{v}^c) \cdot \boldsymbol{n}]\ \text{d}S \\
&= \int_{S_c(t)} (-p\boldsymbol{n} + \boldsymbol{\tau}' \cdot \boldsymbol{n}) \cdot \boldsymbol{v}\ \text{d}S + \int_{V_c(t)} p\nabla \cdot \boldsymbol{v}\ \text{d}V - \int_{V_c(t)} \phi_v\ \text{d}V \\
&+ \int_{V_c(t)} \rho \boldsymbol{f}_m \cdot \boldsymbol{v}\ \text{d}V
\end{aligned}}$$

$$\tag{6.43}$$

Physically, this equation states that the variation of mechanical energy per unit time in $V_c(t)$ equals the inflow of mechanical energy across the control surface, minus the outflow flux of mechanical energy, plus the power supplied by the external forces $\dot{W}_{\text{ext}}$, plus the expansion power $\dot{W}_{\text{exp}}$, minus the energy lost per unit time due to the viscous dissipation D_v,

$$\frac{\mathrm{d}}{\mathrm{d}t} E^{\text{mec}}(V_c) = \dot{E}^{\text{mec}}_{\text{in}} - \dot{E}^{\text{mec}}_{\text{out}} + \dot{W}_{\text{ext}} + \dot{W}_{\text{exp}} - D_v \qquad (6.44)$$

where

$$D_v = \int_{V_c(t)} \phi_v \ \mathrm{d}V \qquad (6.45)$$

If the body force stems from a stationary potential,

$$\begin{aligned}
\rho \boldsymbol{f}_m \cdot \boldsymbol{v} &= \rho \left(-\nabla U\right) \cdot \boldsymbol{v} \\
&= -\rho \left(\frac{\partial U}{\partial t} + \boldsymbol{v} \cdot \nabla U\right) \\
&= -\left(\frac{\partial \rho U}{\partial t} + \nabla \cdot (\boldsymbol{v}\rho U)\right)
\end{aligned} \qquad (6.46)$$

Employing the potential, the equation of mechanical energy can be written as

$$\frac{\partial(\rho \frac{1}{2}v^2 + \rho U)}{\partial t} + \nabla \cdot (\rho \boldsymbol{v} \frac{1}{2} v^2 + \rho \boldsymbol{v} U) = -\nabla \cdot (p\boldsymbol{v}) + \nabla \cdot (\boldsymbol{\tau}' \boldsymbol{v}) + p\nabla \cdot \boldsymbol{v} - \phi_v \quad (6.47)$$

and the integral counterpart is

$$\boxed{\begin{aligned}
\frac{\mathrm{d}}{\mathrm{d}t} \int_{V_c(t)} \rho \left(\tfrac{1}{2}v^2 + U\right) \ \mathrm{d}V &+ \int_{S_c(t)} \rho \left(\frac{1}{2}v^2 + U\right) \left[(\boldsymbol{v} - \boldsymbol{v}^c) \cdot \boldsymbol{n}\right] \ \mathrm{d}S \\
&= \int_{S_c(t)} (-p\boldsymbol{n} + \boldsymbol{\tau}' \cdot \boldsymbol{n}) \cdot \boldsymbol{v} \ \mathrm{d}S + \int_{V_c(t)} p\nabla \cdot \boldsymbol{v} \ \mathrm{d}V - \int_{V_c(t)} \phi_v \ \mathrm{d}V
\end{aligned}}$$
$$(6.48)$$

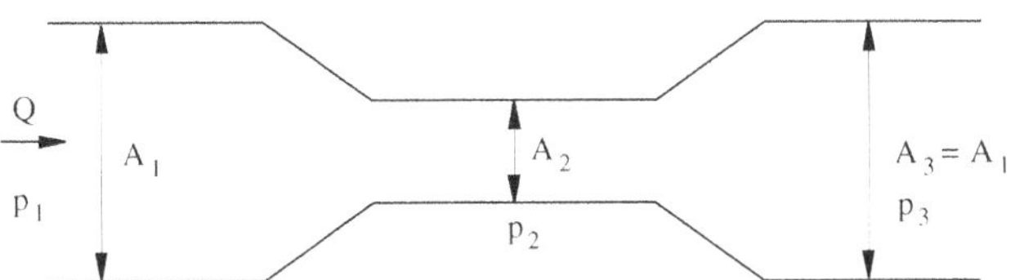

Fig. 6.6. This converging-diverging duct is called a Venturi and can be used as a flowmeter.

Example 6.7 (Venturi meter). A Venturi is a converging-diverging nozzle that can be used as a flowmeter (see Fig. 6.6). In order to measure the flow rate, there are two pressure gages, one at the duct and the other at the throat. Assuming ideal flow (i.e. neglecting the viscous dissipation) determine the relation between the pressure difference at the gages, $\Delta p = p_1 - p_2$, and the volumetric flow rate Q.

Solution. The mass conservation equation reads

$$\underbrace{\frac{\mathrm{d}}{\mathrm{d}t} \int_{V_c(t)} \rho \,\mathrm{d}V}_{\text{steady}} + \int_{S_1+S_2+S_p} \rho \left[(\boldsymbol{v} - \boldsymbol{v}^c) \cdot \boldsymbol{n}\right] \mathrm{d}S = 0$$

and simplifies for uniform variables to

$$Q = v_1 A_1 = v_2 A_2$$

The mechanical energy equation implies

$$\underbrace{\frac{\mathrm{d}}{\mathrm{d}t} \int_{V_c(t)} \rho \left(\frac{1}{2} v^2 + U\right) \mathrm{d}V}_{\text{steady}} + \underbrace{\int_{S_1+S_2+S_p} \rho \left(\frac{1}{2} v^2 + U\right) \left[(\boldsymbol{v} - \boldsymbol{v}^c) \cdot \boldsymbol{n}\right] \mathrm{d}S}_{U=\text{const (horizontal Venturi)}} =$$

$$= \int_{S_c(t)} (-p\boldsymbol{n} + \boldsymbol{\tau}' \cdot \boldsymbol{n}) \cdot \boldsymbol{v} \,\mathrm{d}S$$

$$+ \underbrace{\int_{V_c(t)} p\, \boldsymbol{\nabla} \cdot \boldsymbol{v} \,\mathrm{d}V}_{\rho=\text{const (incompressible flow)}}$$

$$\underbrace{- \int_{V_c(t)} \phi_v \,\mathrm{d}V}_{\text{problem hypothesis}}$$

The non-vanishing integrals are

$$\int_{S_1+S_2} \rho \left(\frac{1}{2} v^2\right) (\boldsymbol{v} \cdot \boldsymbol{n}) \,\mathrm{d}S = \rho(\tfrac{1}{2} v_1^2)(-Q) + \rho(\tfrac{1}{2} v_2^2)(+Q)$$

$$\int_{S_1+S_2+S_p} -p\boldsymbol{n} \cdot \boldsymbol{v} \,\mathrm{d}S = p_1 v_1 A_1 - p_2 v_2 A_2 + \underbrace{\int_{S_p} -p\boldsymbol{n} \cdot \boldsymbol{v} \,\mathrm{d}S}_{v \perp n}$$

$$= Q(p_1 - p_2)$$

$$\int_{S_1+S_2+S_p} (\boldsymbol{\tau}' \cdot \boldsymbol{n}) \cdot \boldsymbol{v} \,\mathrm{d}S = \underbrace{\int_{S_1+S_2} (\boldsymbol{\tau}' \cdot \boldsymbol{n}) \cdot \boldsymbol{v} \,\mathrm{d}S}_{\tau' \approx 0 \text{ inlet, outlet}} + \underbrace{\int_{S_p} (\boldsymbol{\tau}' \cdot \boldsymbol{n}) \cdot \boldsymbol{v} \,\mathrm{d}S}_{\text{ideal flow}}$$

Gathering all the terms,

$$\rho Q\left(-\frac{1}{2}v_1^2 + \frac{1}{2}v_2^2\right) = Q(p_1 - p_2)$$

from where the result we are looking for is

$$Q = A_1 \sqrt{\frac{2(p_1 - p_2)}{\rho\left[\left(\frac{A_1}{A_2}\right)^2 - 1\right]}}$$

Remark 6.3. In order to take into account the energy losses, an experimental factor is included in the above formula, so the real volumetric flow rate is

$$Q = A_1\eta \sqrt{\frac{2(p_1 - p_2)}{\rho\left[\left(\frac{A_1}{A_2}\right)^2 - 1\right]}}$$

where $0 < \eta < 1$, lies between $0.98 - 0.99$.

Example 6.8 (Viscous dissipation in a burner). In the burner of the above example, assuming that the flow is incompressible, calculate the energy dissipated per unit time.

Solution. Let us apply the mechanical energy equation with the usual hypothesis,

$$\underbrace{\frac{\mathrm{d}}{\mathrm{d}t}\int_{V_c(t)} \rho\left(\frac{1}{2}v^2 + U\right)\mathrm{d}V}_{\text{steady}} + \int_{S_c+S_s+S_p} \rho\left(\frac{1}{2}v^2 + U\right)\left[(v - v^c)\cdot n\right]\mathrm{d}S$$

$$= \int_{S_c(t)} (-p n + \tau'\cdot n)\cdot v\,\mathrm{d}S$$

$$\underbrace{+ \int_{V_c(t)} p\, \nabla\cdot v\,\mathrm{d}V}_{\rho=\text{const (incompressible flow)}}$$

$$\underbrace{- \int_{V_c(t)} \phi_v\,\mathrm{d}V}_{\text{viscous dissipation in } V_c(t)=-D_v}$$

In particular, the convective term and the power due to surface forces are

$$\int_{S_c+S_s} \rho \left(\frac{1}{2} v^2 + U \right) (\boldsymbol{v} \cdot \boldsymbol{n}) \, \mathrm{d}S = -\rho(\tfrac{1}{2}v_1^2 + gh_1)v_1 S_1 + \rho(\tfrac{1}{2}v_2^2 + gh_2)v_2 S_2$$

$$\int_{S_1+S_2+S_p} -p\boldsymbol{n} \cdot \boldsymbol{v} \, \mathrm{d}S = p_1 v_1 S_1 - p_2 v_2 S_2 + \underbrace{\int_{S_p} -p\boldsymbol{n} \cdot \boldsymbol{v} \, \mathrm{d}S}_{\boldsymbol{v}=\boldsymbol{0}}$$

$$\int_{S_1+S_2+S_p} (\boldsymbol{\tau}' \cdot \boldsymbol{n}) \cdot \boldsymbol{v} \, \mathrm{d}S = \underbrace{\int_{S_1+S_2} (\boldsymbol{\tau}' \cdot \boldsymbol{n}) \cdot \boldsymbol{v} \, \mathrm{d}S}_{\boldsymbol{\tau}'\approx\boldsymbol{0}\,\text{inlet, outlet}} + \underbrace{\int_{S_p} (\boldsymbol{\tau}' \cdot \boldsymbol{n}) \cdot \boldsymbol{v} \, \mathrm{d}S}_{\boldsymbol{v}=\boldsymbol{0}\,\text{at walls if }\mu\neq 0}$$

Operating,

$$
\begin{aligned}
D_v &= \rho(\tfrac{1}{2}v_1^2 + gh_1)v_1 S_1 - \rho(\tfrac{1}{2}v_2^2 + gh_2)v_2 S_2 + p_1 v_1 S_1 - p_2 v_2 S_2 \\
&= \rho Q \left[(\tfrac{1}{2}v_1^2 + gh_1 + \tfrac{p_1}{\rho}) - (\tfrac{1}{2}v_2^2 + gh_2 + \tfrac{p_2}{\rho}) \right]
\end{aligned}
$$

Therefore, since the flow is incompressible, the viscous dissipation in the control volume equals the mechanical energy difference (including the pressure) between the inlet and the outlet.

The Bernoulli Equation

A very important version of the mechanical energy equation is the Bernoulli equation. The simplest form of this equation states that for stationary, incompressible, ideal flow, the mechanical energy plus the pressure is constant along a streamline if there is no power supplied to the flow,

$$\boxed{\frac{1}{2}\rho v^2 + p + \rho U = C_{\mathrm{sl}}} \tag{6.49}$$

where the constant C_{sl} depends on the streamline.

Example 6.9 (Venturi meter). Let us repeat the example of the Venturi meter employing the Bernoulli equation (see Example 6.7). For that purpose, take the centerline streamline between 1 and 2. The potential energy of that streamline is constant because it is horizontal and, neglecting the viscous losses, the Bernoulli equation for a liquid can be written as

$$\frac{1}{2}\rho v_1^2 + p_1 = \frac{1}{2}\rho v_2^2 + p_2$$

This is the same expression obtained in Example 6.7 for the mechanical energy equation.

6.5.2 Internal Energy Equation

Subtracting the mechanical energy equation from the total energy equation results in the internal energy equation,

$$\frac{\mathrm{d}}{\mathrm{d}t} \int_{V_c(t)} \rho e \,\mathrm{d}V + \int_{S_c(t)} \rho e \left[(\boldsymbol{v} - \boldsymbol{v}^c) \cdot \boldsymbol{n}\right] \mathrm{d}S$$
$$= - \int_{S_c(t)} p \boldsymbol{\nabla} \cdot \boldsymbol{v} \,\mathrm{d}V + \int_{V_c(t)} \phi_v \,\mathrm{d}V - \int_{S_c(t)} \boldsymbol{q} \cdot \boldsymbol{n} \,\mathrm{d}S + \int_{V_c(t)} \dot{q}_v \,\mathrm{d}V$$

$$(6.50)$$

This equation shows that the rate of variation of the internal energy in $V_c(t)$ $\frac{\mathrm{d}}{\mathrm{d}t} E^{\mathrm{int}}(V_c)$ equals the ingoing flux of internal energy through the control surface, minus the outgoing internal energy flux, minus the expansion power $\dot{W}_{\mathrm{exp}}$ plus the power lost due to the viscous dissipation D_v plus the heat added to the control volume,

$$\frac{\mathrm{d}}{\mathrm{d}t} E^{\mathrm{int}}(V_c) = \dot{E}^{\mathrm{int}}_{\mathrm{in}} - \dot{E}^{\mathrm{int}}_{\mathrm{out}} - \dot{W}_{\mathrm{exp}} + D_v + \dot{Q}_{\mathrm{in}} \qquad (6.51)$$

Example 6.10 (Viscous dissipation in a burner). In the example of the burner, calculate the viscous dissipation in the control volume using the internal energy equation and verify that the total energy equation is satisfied.
Solution.
(a) First, let us apply the internal energy equation using the standard hypothesis.

$$\frac{\mathrm{d}}{\mathrm{d}t} \underbrace{\int_{V_c(t)} \rho e \,\mathrm{d}V}_{\text{steady}} + \int_{S_c+S_s+S_p} \rho e \left[(\boldsymbol{v} - \boldsymbol{v}^c) \cdot \boldsymbol{n}\right] \mathrm{d}S$$

$$= - \underbrace{\int_{S_c(t)} \boldsymbol{q} \cdot \boldsymbol{n} \,\mathrm{d}S}_{\dot{Q}_{\mathrm{in}} - \dot{Q}_{\mathrm{out}}} + \underbrace{\int_{V_c(t)} \dot{q}_v \,\mathrm{d}V}_{\text{absent, considered in } \boldsymbol{q}}$$

$$- \underbrace{\int_{V_c(t)} p \boldsymbol{\nabla} \cdot \boldsymbol{v} \,\mathrm{d}V}_{\rho = \text{const. (incompressible flow)}}$$

$$+ \underbrace{\int_{V_c(t)} \phi_v \,\mathrm{d}V}_{\text{viscous dissipation in} V_c(t) = D_v}$$

In particular, the convective term is

$$\int_{S_c+S_s} \rho e \,(\boldsymbol{v} \cdot \boldsymbol{n}) \,\mathrm{d}S = -\rho e_1 v_1 S_1 + \rho e_2 v_2 S_2$$

$$= \rho Q (e_2 - e_1)$$

Operating,

$$D_v = \rho Q(e_2 - e_1) - (\dot{Q}_{\text{in}} - \dot{Q}_{\text{out}})$$

(b) Substituting D_v into the mechanical energy equation yields

$$\rho Q \left[(e_2 + \tfrac{1}{2}v_2^2 + gh_2 + \tfrac{p_2}{\rho}) - (e_1 + \tfrac{1}{2}v_1^2 + gh_1 + \tfrac{p_1}{\rho}) \right] = \dot{Q}_{\text{in}} - \dot{Q}_{\text{out}}$$

which is an expression of the total energy equation.

6.5.3 Energy Transfer Between Mechanical and Internal Energy

In the mechanical and internal energy equations there are two common terms, the expansion power $\dot{W}_{\text{exp}}$ and the viscous dissipation D_v.

The *expansion power*, when positive, increases the mechanical energy at the expense of decreasing the internal energy. The expansion power can also be negative when the fluid particles are compressed, in which case, the mechanical energy is used to increase the internal energy. Therefore, this type of power is *reversible*, it can be transformed back and forth into mechanical and internal energy.

However, the *viscous dissipation* is always positive. It appears with a negative sign in the mechanical energy equation whereas with a positive sign in the internal energy equation. Since the viscous dissipation is always positive, this power transfer has only one direction: it always decreases the mechanical energy to increase the internal energy. Thus, this is an *irreversible* energy transfer. As a consequence, friction always tends to decrease the mechanical energy, which is transformed into internal energy, increasing the fluid temperature.

Note that none of the above process changes the total energy of the fluid, because they are exchanging energy of two different types (mechanical and internal energy). This is why they do not appear in the total energy equation.

6.6 Conservation of Chemical Species Equation

6.6.1 Introductory Definitions

Let us assume that the fluid is a mixture of n_{esp} chemical species. Each chemical species will be designated by the index A, $1 \leq A \leq n_{\text{esp}}$. Next, mass and molar measures of the chemical concentration are explained.

Mass measures of chemical concentration

Definition 6.1 (Mass concentration). *The mass concentration of the chemical species A, ρ_A, is the mass of the chemical species A per unit volume of mixture. The unit in the SI is* kg/m^3.

Definition 6.2 (Mass fraction). *The mass fraction of the chemical species A, Y_A, is the mass of the chemical species A present per unit mass of mixture. Note $0 \leq Y_A \leq 1$. The mass fraction is dimensionless.*

Note that between both measures the following relations hold

$$
\begin{aligned}
1 &= \sum_{A=1}^{n_{esp}} Y_A \\
\rho &= \sum_{A=1}^{n_{esp}} \rho_A \\
\rho_A &= \rho Y_A
\end{aligned}
\tag{6.52}
$$

Molar measures of chemical concentration

Let us recall that one mole (units mol) of substance corresponds to an amount equal to 6.023×10^{23} molecules. This constant is called the Avogadro constant.

Definition 6.3 (Molar or mole mixture concentration). *The molar concentration of a mixture, c, is the number of moles of mixture per unit volume of mixture. The units in the SI are the $\mathrm{mol/m}^3$ but $\mathrm{mol/l}$ is used frequently.*

Definition 6.4 (Molar or mole concentration). *The molar concentration of the chemical species A, c_A, is the number of moles of the chemical species A per unit of mixture volume. The units in the SI are the $\mathrm{mol/m}^3$ but $\mathrm{mol/l}$ is used frequently.*

Definition 6.5 (Molar fraction). *The molar fraction of the chemical species A, X_A, is the number of moles of the chemical species A per number of total moles of mixture. Note $0 \leq X_A \leq 1$. The molar fraction is dimensionless.*

The molar concentrations obey the following relations:

$$
\begin{aligned}
1 &= \sum_{A=1}^{n_{esp}} X_A \\
c &= \sum_{A=1}^{n_{esp}} c_A \\
c_A &= c x_A
\end{aligned}
\tag{6.53}
$$

Relations between mass and molar concentrations

Let us recall that the *molar mass* M_A represents the mass of one mole of substance. Its units are g/mol or kg/kmol.

The following relations between mass and molar concentrations hold

$$
\begin{aligned}
\rho_A &= M_A c_A \\
\rho &= M c
\end{aligned}
\tag{6.54}
$$

where M_A is the *molar mass* (or molecular weight) of the chemical species A and M the molar mass of the mixture

$$M \;=\; \sum_{A=1}^{n_{esp}} X_A M_A$$
$$\;=\; \left(\sum_{A=1}^{n_{esp}} Y_A/M_A\right)^{-1}$$

(6.55)

As far as fraction concentration measures are concerned,

$$\boxed{\begin{aligned} Y_A &= \frac{X_A M_A}{\sum_{B=1}^{n_{esp}} X_B M_B} \\[2mm] X_A &= \frac{Y_A/M_A}{\sum_{B=1}^{n_{esp}} X_B/M_B} \end{aligned}}$$

(6.56)

Average fluid velocity of the mixture

Because each chemical species may travel at a different speed $\boldsymbol{v}_A$, we need to introduce a definition of the average fluid velocity $\boldsymbol{v}$.

Definition 6.6 (Chemical species velocity). *We will denote with $\boldsymbol{v}_A$ the velocity of the chemical species A.*

There are several possibilities for defining the average fluid velocity. Here, we will consider mass and molar averaging.

Definition 6.7 (Mass average fluid velocity). *We will denote by $\boldsymbol{v}$ the mass mean velocity of the mixture,*

$$\boldsymbol{v} = \frac{\sum_{A=1}^{n_{esp}} \rho_A \boldsymbol{v}_A}{\rho} = \sum_{A=1}^{n_{esp}} Y_A \boldsymbol{v}_A$$

(6.57)

In general $\boldsymbol{v} \neq \boldsymbol{v}_A$.

Definition 6.8 (Molar average fluid velocity). *We can define a mean molar velocity of the mixture as*

$$\boldsymbol{v}^m = \frac{\sum_{A=1}^{n_{esp}} c_A \boldsymbol{v}_A}{c} = \sum_{A=1}^{n_{esp}} X_A \boldsymbol{v}_A$$

(6.58)

6.6.2 Derivation of the Conservation Equations

Since every chemical species travels at a different speed $\boldsymbol{v}_A$, it is complicated to derive the chemical species mass conservation equation for a fluid volume, which moves at one of the mean velocities defined above.

Instead, we will take a specific control volume for each species. In particular, for the species A let us select the control volume $V_{cA}(t)$ that contains always the same particles of that species. In absence of chemical reaction, the mass of species A in the control volume $V_{cA}(t)$ is constant,

$$\frac{\mathrm{d}}{\mathrm{d}t}\int_{V_{cA}(t)} \rho_A \; \mathrm{d}V = 0 \tag{6.59}$$

In the presence of chemical reactions, there will be a global rate of production per unit volume for every chemical species $\dot{\omega}_A$ $[\mathrm{kg/(m^3\ s)}]$ and finally the mass conservation equation can be written as

$$\boxed{\frac{\mathrm{d}}{\mathrm{d}t}\int_{V_{cA}(t)} \rho_A \; \mathrm{d}V = \int_{V_{cA}(t)} \dot{\omega}_A \; \mathrm{d}V} \tag{6.60}$$

It is more practical to rewrite the above equation for an arbitrary control volume. Applying the transport theorems to a control volume $V_c(t)$ that coincides at time t with $V_{cA}(t)$,

$$V_c(t) = V_{cA}(t) \tag{6.61}$$

we have

$$\begin{aligned}
\frac{\mathrm{d}}{\mathrm{d}t}\int_{V_{cA}(t)} \rho_A \; \mathrm{d}V &= \int_{V_{cA}(t)} \frac{\partial \rho_A}{\partial t} \; \mathrm{d}V + \int_{S_{cA}(t)} \rho_A \boldsymbol{v}_A \cdot \boldsymbol{n} \; \mathrm{d}S \\
\frac{\mathrm{d}}{\mathrm{d}t}\int_{V_c(t)} \rho_A \; \mathrm{d}V &= \int_{V_c(t)} \frac{\partial \rho_A}{\partial t} \; \mathrm{d}V + \int_{S_c(t)} \rho_A \boldsymbol{v}^c \cdot \boldsymbol{n} \; \mathrm{d}S
\end{aligned} \tag{6.62}$$

Subtracting the second equation from the first one, and subtracting and adding the surface integral

$$\begin{aligned}
\frac{\mathrm{d}}{\mathrm{d}t}\int_{V_{cA}(t)} \rho_A \; \mathrm{d}V &= \frac{\mathrm{d}}{\mathrm{d}t}\int_{V_c(t)} \rho_A \; \mathrm{d}V + \int_{S_c(t)} \rho_A(\boldsymbol{v}_A - \boldsymbol{v}^c) \cdot \boldsymbol{n} \; \mathrm{d}S \\
&\quad - \int_{S_{cA}(t)} \rho_A \boldsymbol{v} \cdot \boldsymbol{n} \; \mathrm{d}S + \int_{S_{cA}(t)} \rho_A \boldsymbol{v} \cdot \boldsymbol{n} \; \mathrm{d}S \\
&= \frac{\mathrm{d}}{\mathrm{d}t}\int_{V_c(t)} \rho_A \; \mathrm{d}V + \int_{S_c(t)} \rho_A(\boldsymbol{v} - \boldsymbol{v}^c) \cdot \boldsymbol{n} \; \mathrm{d}S \\
&\quad + \int_{S_c(t)} \rho_A(\boldsymbol{v}_A - \boldsymbol{v}) \cdot \boldsymbol{n} \; \mathrm{d}S
\end{aligned} \tag{6.63}$$

Substituting and reorganizing,

$$\boxed{\begin{aligned}
\frac{\mathrm{d}}{\mathrm{d}t}\int_{V_c(t)} \rho_A \; \mathrm{d}V &+ \int_{S_c(t)} \rho_A \left[(\boldsymbol{v} - \boldsymbol{v}^c) \cdot \boldsymbol{n}\right] \; \mathrm{d}S \\
&= -\int_{S_c(t)} \rho_A(\boldsymbol{v}_A - \boldsymbol{v}) \cdot \boldsymbol{n} \; \mathrm{d}S + \int_{V_c(t)} \dot{\omega}_A \; \mathrm{d}V
\end{aligned}} \tag{6.64}$$

The physical interpretation of the above equation is as follows. The increase of mass of the chemical species A per unit time in the control volume $\frac{\mathrm{d}}{\mathrm{d}t}M_A(V_c)$

equals the ingoing mass flux of chemical species A through the control surface $\dot{M}_{A\,in} - \dot{M}_{A\,out}$ plus the inward diffusion transport $J_{A\,in} - J_{A\,out}$ [kg/s] and the production due to chemical reaction, $\dot{G}_A$:

$$\frac{\mathrm{d}}{\mathrm{d}t} M_A(V_c) = (\dot{M}_{A\,in} - \dot{M}_{A\,out}) + (J_{A\,in} - J_{A\,out}) + \dot{G}_A \tag{6.65}$$

Remark 6.4. The first integral on the right-hand side contains the velocity difference between the chemical species $\boldsymbol{v}_A$ and the mean flow velocity $\boldsymbol{v}$. This is modeled by molecular diffusion of the species A, and is written as a *diffusion mass flux* vector, $\boldsymbol{j}_A$ [kg/(m^2 s)]

$$\boxed{\boldsymbol{j}_A = \rho_A(\boldsymbol{v}_A - \boldsymbol{v}) \qquad (\text{no sum})} \tag{6.66}$$

such that

$$-\int_{S_c(t)} \rho_A(\boldsymbol{v}_A - \boldsymbol{v}) \cdot \boldsymbol{n} \ \mathrm{d}S = -\int_{S_c(t)} \boldsymbol{j}_A \cdot \boldsymbol{n} \ \mathrm{d}S \tag{6.67}$$

Remark 6.5. Many times, instead of the diffusion mass flux, the *diffusion velocity* is imployed,

$$\boldsymbol{V}_A = \frac{\boldsymbol{j}_A}{\rho_A} = (\boldsymbol{v}_A - \boldsymbol{v}) \qquad (\text{no sum}) \tag{6.68}$$

Remark 6.6. Not all the chemical conservation equations are independent. In fact, its sum adds up to the mass conservation equation. Let us check this term by term.

$$\sum_{A=1}^{n_{\mathrm{esp}}} \frac{\mathrm{d}}{\mathrm{d}t} \int_{V_c(t)} \rho_A \ \mathrm{d}V = \frac{\mathrm{d}}{\mathrm{d}t} \int_{V_c(t)} \sum_{A=1}^{n_{\mathrm{esp}}} \rho_A \ \mathrm{d}V$$

$$= \frac{\mathrm{d}}{\mathrm{d}t} \int_{V_c(t)} \rho \ \mathrm{d}V \tag{6.69}$$

$$\sum_{A=1}^{n_{\mathrm{esp}}} \int_{S_c(t)} \rho_A \left[(\boldsymbol{v} - \boldsymbol{v}^c) \cdot \boldsymbol{n} \right] \mathrm{d}S = \int_{S_c(t)} \sum_{A=1}^{n_{\mathrm{esp}}} \rho_A \left[(\boldsymbol{v} - \boldsymbol{v}^c) \cdot \boldsymbol{n} \right] \mathrm{d}S$$

$$= \int_{S_c(t)} \rho \left[(\boldsymbol{v} - \boldsymbol{v}^c) \cdot \boldsymbol{n} \right] \mathrm{d}S \tag{6.70}$$

$$\sum_{A=1}^{n_{\mathrm{esp}}} \int_{S_c(t)} \rho_A(\boldsymbol{v}_A - \boldsymbol{v}) \cdot \boldsymbol{n} \ \mathrm{d}S = \int_{S_c(t)} \sum_{A=1}^{n_{\mathrm{esp}}} \rho_A(\boldsymbol{v}_A - \boldsymbol{v}) \cdot \boldsymbol{n} \ \mathrm{d}S$$

$$= \int_{S_c(t)} \left(\sum_{A=1}^{n_{\mathrm{esp}}} \rho_A \boldsymbol{v}_A - \sum_{A=1}^{n_{\mathrm{esp}}} \rho_A \boldsymbol{v} \right) \cdot \boldsymbol{n} \ \mathrm{d}S$$

$$= \int_{S_c(t)} (\rho \boldsymbol{v} - \rho \boldsymbol{v}) \cdot \boldsymbol{n} \ \mathrm{d}S = 0 \tag{6.71}$$

$$\sum_{A=1}^{n_{\mathrm{esp}}} \int_{V_c(t)} \dot{\omega}_A \ \mathrm{d}V = \int_{V_c(t)} \sum_{A=1}^{n_{\mathrm{esp}}} \dot{\omega}_A \ \mathrm{d}V = 0 \tag{6.72}$$

Note that

$$\sum_{A=1}^{n_{\mathrm{esp}}} \boldsymbol{j}_A = 0 \tag{6.73}$$

$$\sum_{A=1}^{n_{\mathrm{esp}}} \dot{\omega}_A = 0 \tag{6.74}$$

Remark 6.7. See Appendix I for how to compute $\dot{\omega}_A$.

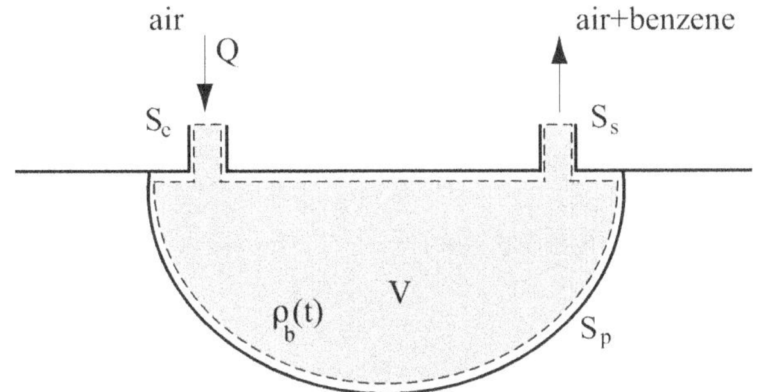

Fig. 6.7. Soil cleaning.

Example 6.11 (Soil cleaning). There is a soil polluted with an initial concentration of benzene $\rho_b(t = 0) = \rho_{b0}$. In order to clean the soil, air is introduced into the ground through an inlet. The fluid mixes with the benzene and the mixture is extracted at the outlet. Determine the evolution of mean benzene concentration $\rho_b(t)$ as a function of time t, the volumetric flow rate of air Q and the volume of treated soil V.

Solution. Let us apply the chemical species conservation equation to the benzene inside the control volume, which encompasses the polluted area of volume V. Any chemical reaction and the molecular diffusion will be ignored. Also, the concentration of benzene in the control volume will be assumed uniform, i.e. $\rho_b(t)$. Thus,

$$\frac{\mathrm{d}}{\mathrm{d}t} \int_{V_c(t)} \rho_b \, \mathrm{d}V + \int_{S_c(t)} \rho_b \left[(\boldsymbol{v} - \boldsymbol{v}^c) \cdot \boldsymbol{n} \right] \mathrm{d}S = \underbrace{- \int_{S_c(t)} \rho_b (\boldsymbol{v}_b - \boldsymbol{v}) \cdot \boldsymbol{n} \, \mathrm{d}S}_{\text{well mixed}}$$

$$+ \underbrace{\int_{V_c(t)} \dot{\omega}_b \, \mathrm{d}V}_{\text{no chemical reaction}}$$

The temporal variation of the mass of benzene in the control volume is

$$\frac{\mathrm{d}}{\mathrm{d}t} \int_{V_c(t)} \rho_b(t) \, \mathrm{d}V = \frac{\mathrm{d}\rho_b(t)}{\mathrm{d}t} V$$

There exists only mass flow across the inlet and outlet, S_e and S_s, so the flow integral becomes

$$\int_{S_e + S_p + S_s} \rho_b \left[(\boldsymbol{v} - \boldsymbol{v}^c) \cdot \boldsymbol{n} \right] \mathrm{d}S = -\rho_{be} Q + \rho_b Q$$

The concentration of benzene at the inlet is zero, $\rho_{be} = 0$, since it is fresh air. Furthermore, let us take the concentration of benzene at the outlet equal to the mean concentration of benzene at the polluted area, i.e. $\rho_{bs} = \rho_b$. Gathering all the components, we arrive at the ordinary differential equation

$$\frac{\mathrm{d}\rho_b}{\mathrm{d}t} = -\frac{\rho_b Q}{V}$$

Integrating in time and imposing the initial condition,

$$\rho_b(t) = \rho_{b0} \, e^{-\frac{Qt}{V}}$$

from where it is concluded that the pollution disappears as an exponentially decreasing function. At the beginning, the soil gets cleaned very rapidly, but the cleaner it is, the slower the process. Thus, it is more difficult to eliminate small than large concentrations.

6.6.3 Chemical Species Equations for Molar Concentrations

It is possible to express the conservation equation of chemical species in molar concentrations. To do so, equation (6.64) is divided by the molecular mass of the corresponding chemical species, M_A. The rate of mass production transforms into the rate of molar production [mol/(m^3 s)], i.e.

$$\dot{\omega}'_A = \frac{\dot{\omega}_A}{M_A} \tag{6.75}$$

The equation of the evolution of the molar concentration with respect to the mass mean velocity is

$$\boxed{\begin{aligned} \frac{\mathrm{d}}{\mathrm{d}t} \int_{V_c(t)} c_A \, \mathrm{d}V + \int_{S_c(t)} c_A \left[(\boldsymbol{v} - \boldsymbol{v}^c) \cdot \boldsymbol{n} \right] \mathrm{d}S \\ = -\int_{S_c(t)} c_A (\boldsymbol{v}_A - \boldsymbol{v}) \cdot \boldsymbol{n} \, \mathrm{d}S + \int_{V_c(t)} \dot{\omega}'_A \, \mathrm{d}V \end{aligned}} \tag{6.76}$$

The molecular diffusivity can be rewritten as a function of the vector of molar flux $\boldsymbol{j}'_A$ [mol/(m^2 s)],

$$\boldsymbol{j}'_A = c_A (\boldsymbol{v}_A - \boldsymbol{v}) \qquad \text{(no sum)} \tag{6.77}$$

6.6.4 Equations with Respect to the Molar Average Velocity

The above chemical species equations have been written with respect to the mass average mixture velocity. If instead, the molar average velocity v^m is employed, the corresponding equations are obtained:

$$\frac{\mathrm{d}}{\mathrm{d}t} \int_{V_c(t)} \rho_A \, \mathrm{d}V + \int_{S_c(t)} \rho_A \left[(v^m - v^c) \cdot n \right] \mathrm{d}S$$
$$= - \int_{S_c(t)} \rho_A (v_A - v^m) \cdot n \, \mathrm{d}S + \int_{V_c(t)} \dot{\omega}_A \, \mathrm{d}V \tag{6.78}$$

Apparently, everything looks similar, but the mass diffusion flux $[\mathrm{kg}/(\mathrm{m}^3 \ \mathrm{s})]$ is written with respect to the molar averaged velocity,

$$j_A^m = \rho_A (v_A - v^m) \qquad \text{(no sum)}$$

For molar concentrations,

$$\frac{\mathrm{d}}{\mathrm{d}t} \int_{V_c(t)} c_A \, \mathrm{d}V + \int_{S_c(t)} c_A \left[(v^m - v^c) \cdot n \right] \mathrm{d}S =$$
$$= - \int_{S_c(t)} c_A (v_A - v^m) \cdot n \, \mathrm{d}S + \int_{V_c(t)} \dot{\omega}'_A \, \mathrm{d}V \tag{6.79}$$

where the molar diffusion flux $[\mathrm{mol}/(\mathrm{m}^2 \ \mathrm{s})]$ is

$$j_A^{m'} = c_A (v_A - v^m) \qquad \text{(no sum)} \tag{6.80}$$

6.7 Equation of Volume Conservation for Liquids

When the fluid is a mixture of inmiscible liquids, a similar principle to mass conservation can be derived, but applied to the volume of the fluid. If the fluid volume contains several liquids, in the absence of chemical reactions, then its volume is constant and we can write

$$\text{Volume}(V_f) = \text{constant} \tag{6.81}$$

Thus,

$$\frac{\mathrm{d}}{\mathrm{d}t} \text{Volume}(V_f) = 0 \tag{6.82}$$

The volume of V_f is simply

$$\text{Volume}(V_f) = \int_{V_f(t)} \mathrm{d}V \tag{6.83}$$

Consequently, the principle of volume conservation for liquids is

$$\boxed{\frac{\mathrm{d}}{\mathrm{d}t} \int_{V_f(t)} \mathrm{d}V = 0}$$

(6.84)

Applying the third transport theorem with $\phi = 1$,

$$\boxed{\frac{\mathrm{d}}{\mathrm{d}t} \int_{V_c(t)} \mathrm{d}V + \int_{S_c(t)} [(\boldsymbol{v} - \boldsymbol{v}^c) \cdot \boldsymbol{n}] \, \mathrm{d}S = 0}$$

(6.85)

The physical interpretation of this equation is similar to that of other equations. All the fluid volume that enters in the control volume has to exit or increase the volume $V_c(t)$.

Remark 6.8. If there is only one liquid, this equation is equivalent to the mass conservation equation.

Remark 6.9. If there are two or more liquids with different densities, this equation is linearly independent of the mass conservation equation.

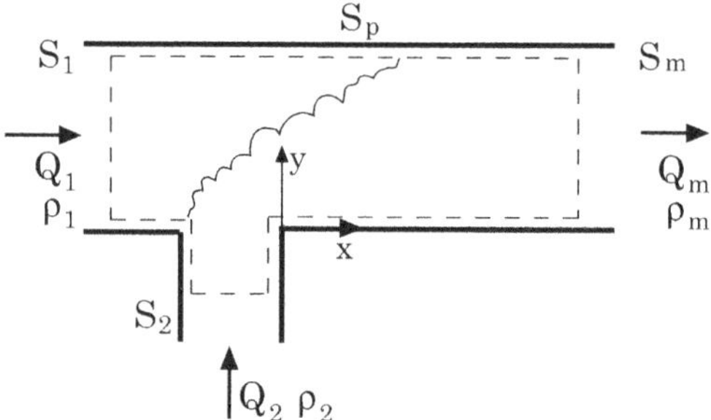

Fig. 6.8. Mixing process through perpendicular ducts.

Example 6.12 (Mixing process through perpendicular jets). A method employed in the chemical industry to mix two fluids consists of merging two flows perpendicularly. If the flow is stationary, using the data of the Figure for the inlet liquid streams, determine the density and volumetric flow rate at the exit.

Solution. Let us take the control volume of Fig. 6.8. From the mass conservation equation,

$$\underbrace{\frac{\mathrm{d}}{\mathrm{d}t} \int_{V_c(t)} \rho \, \mathrm{d}V}_{\text{stationary}} + \int_{S_1+S_2+S_s+S_p} \rho \, [(\boldsymbol{v} - \boldsymbol{v}^c) \cdot \boldsymbol{n}] \, \mathrm{d}S = 0$$

The integral on S_p vanishes since there is no flow through solid walls. Assuming that the density is uniform at the inlet and outlet sections,

$$-\rho_1 Q_1 - \rho_2 Q_2 + \rho_m Q_m = 0$$

In this equation, there are two unknowns, ρ_m and Q_m. Thus, another equation is necessary. If we used the $n_{\text{esp}} - 1 = 1$ independent equation of chemical species, we would add a new unknown, namely the chemical concentration of the species 1 or 2 at the exit.

Therefore, a different equation is needed, in particular, the volume conservation equation for liquids:

$$\underbrace{\frac{\mathrm{d}}{\mathrm{d}t} \int_{V_c(t)} \mathrm{d}V}_{\text{stationary}} + \int_{S_1+S_2+S_s+S_p} \left[(\boldsymbol{v} - \boldsymbol{v}^c) \cdot \boldsymbol{n}\right] \mathrm{d}S = 0$$

Simplifying

$$-Q_1 - Q_2 + Q_m = 0$$

from where it can be concluded that the exit volumetric flow rate is the sum of the inlet flow rates:

$$Q_m = Q_1 + Q_2$$

Introducing this result into the mass conservation equation,

$$\rho_m = \frac{\rho_1 Q_1 + \rho_2 Q_2}{Q_1 + Q_2}$$

6.8 Outline

It is enlightening to remark that all transport equations share common features, since all of them can be cast in the general form

$$\frac{\mathrm{d}}{\mathrm{d}t} \int_{V_c(t)} \rho\phi \, \mathrm{d}V + \int_{S_c(t)} \rho\phi \left[(\boldsymbol{v} - \boldsymbol{v}^c) \cdot \boldsymbol{n}\right] \mathrm{d}S = \mathcal{D}_\phi + \mathcal{F}_\phi \qquad (6.86)$$

On the left-hand side we can encounter the temporal term and the flux across the control surface (which contains the convective transport). The right-hand side may be zero or contain the diffusion transport $\mathcal{D}_\phi$ and source term $\mathcal{F}_\phi$ depending on the specific equation.

Thus, selecting ϕ properly the various equations can be recovered, as indicated on Table 6.1.

Remark 6.10. Note that $\rho\phi$ is the conserved property density on each equation, that is,

$$\rho\phi = \frac{\text{property}}{\text{volume}}$$

See Table 6.2.

Table 6.1. How to recover the conservation equations from the conserved property per unit mass ϕ.

ϕ	Equation
1	mass conservation
$\boldsymbol{v}$	momentum
$\boldsymbol{r} \times \boldsymbol{v}$	angular momentum
$e + \frac{1}{2}v^2$	total energy conservation
$\frac{1}{2}v^2$	mechanical energy
e	internal energy
Y_A	conservation of chemical species A
x_A/M	idem for molar fluxes
$1/\rho$	volume conservation

Table 6.2. Property density $\rho\phi$ for various conservation equations.

Equation	$\rho\phi$
mass conservation	$\dfrac{\text{mass}}{\text{volume}} = \rho$
momentum	$\dfrac{\text{mass } \boldsymbol{v}}{\text{volume}} = \rho\boldsymbol{v}$
internal energy	$\dfrac{\text{mass } e}{\text{volume}} = \rho e$

Remark 6.11. The convective flux of the property per unit mass ϕ across a surface $S_c(t)$ that moves at a velocity $\boldsymbol{v}^c$ is

$$\int_{S_c(t)} \rho\phi \left[(\boldsymbol{v} - \boldsymbol{v}^c) \cdot \boldsymbol{n} \right] \, \mathrm{d}S$$

which represents the amount of property that flows per unit time across the control surface, i.e.

$$\mathrm{flux}(\rho\phi) = \rho\phi\, Q \;=\; \frac{\text{property}}{\text{volume}} \cdot \frac{\text{volume}}{\text{time}}$$
$$=\; \frac{\text{property}}{\text{time}}$$

6.9 Initial and Boundary Conditions

In order to solve the conservation equations, initial and boundary conditions need to be provided.

6.9.1 Initial Conditions

The initial conditions provide the initial state from which the flow evolves. These data are the values of the fluid variables at the initial state and are relevant in transient calculations.

6.9.2 Boundary Conditions

The boundary conditions are values of flow properties or fluxes that must be provided on the surface of the control volume. The most frequent boundary conditions, used in the examples, are summarized below.

Inlets and Outlets

Typically, at inlets and outlets most of the fluid variables are known, like the velocity v, the temperature T, the pressure p, the chemical concentrations ρ_A, and so on. A good starting point for our analysis can be to assume that these fluid variables are uniform at inlet and outlets.

Due to the difficulty of evaluating τ', generally the viscous stress tensor is neglected, $\tau' \approx 0$. This is only a small approximation if inlets and outlets are chosen in areas of mostly uniform properties. Likewise for the diffusion heat q and the mass transfer fluxes j_A.

Solid Walls – Fluid/Solid interfaces

We can distinguish two cases: ideal and viscous flow. Next, we summarize some boundary conditions in the absence of blowing or suction at the wall (non-porous wall).

Ideal Flow

For an ideal flow, all the diffusivities are zero, i.e. $\mu = \kappa = D_{AB} = 0$ and in this way, the diffusive fluxes vanish, $\tau' = 0$, $q = j_A = 0$ (see Chapter 7). In this case, the following boundary conditions hold. Below, n denotes the unit vector orthogonal to the wall.

(a) The velocity is parallel to the wall. Thus, the velocity component normal to the wall is zero,

$$v_n = v \cdot n = 0 \tag{6.87}$$

This is called the *slip* boundary condition.

(b) For the energy equation, the heat flux vanishes,

$$q_n = \boldsymbol{q} \cdot \boldsymbol{n} = 0 \tag{6.88}$$

(c) Likewise, for the chemical species equations there is no diffusion mass flux,

$$j_{An} = \boldsymbol{j}_A \cdot \boldsymbol{n} = 0 \tag{6.89}$$

Viscous Flow

In real fluids, all the diffusion coefficients are non-zero. In this case, the following boundary conditions hold.

(a) The velocity $\boldsymbol{v}$ of the fluid in contact with a wall equals the wall velocity $\boldsymbol{v}_w$. This is called the *no-slip* boundary condition.
(b) For the energy equation, the fluid temperature at the wall T equals the wall temperature T_w. This boundary condition can be replaced by the heat flux at the wall q_n.
(c) For the chemical species equation, either the concentration ρ_A or the normal mass flux j_{An} can be set. The chemical concentration at the fluid by the wall ρ_{Aw} can be determined from equilibrium data.

Fluid/Fluid Interfaces

Fluid/fluid interfaces separate two fluids with different conditions. These encompass liquid/liquid, liquid/gas and gas/gas interfaces. One important class of interfaces in engineering applications are *free surfaces*, which are liquid/gas interfaces subject to a gravitational field.

The boundary conditions at interfaces are derived by applying the integral conservation equations to a thin control volume that follows the interface. Across the interface it may be important to consider the surface tension σ (Chapter 3).

Let us denote by $\boldsymbol{n}$ and $\boldsymbol{t}$ the normal and tangential directions to the interface, which separates media 1 and 2. The media will be denoted with the superscripts 1 and 2. Next, we present the simplest form of boundary conditions for real fluids when mass transfer and the motion of the interface are neglected.

(a) For real fluids in thermodynamic equilibrium, the tangential fluid velocity and the temperature are continuous across the interface. This is a generalization of the no-slip boundary condition.
(b) The tangential balance of momentum (in direction $\boldsymbol{t}$) for negligible spatial variations of σ yields

$$\tau'^1_{tn} = \tau'^2_{tn} \tag{6.90}$$

That is, the viscous shear stress is continuous across the interface.

In the particular case that the viscosity of one of the media is much smaller than that of the other medium, for example $\mu^1 << \mu^2$, and the fluid is not forced in the media of small viscosity, $u_t^1 \approx 0$, we can approximate

$$\frac{\partial u_t^2}{\partial n} \approx 0 \tag{6.91}$$

This boundary condition can be applied to the liquid side of free surfaces where the gas is almost still.

(c) The normal momentum balance across the interface gives

$$\tau_{nn}^1 - \tau_{nn}^2 = \frac{2\sigma}{R} \tag{6.92}$$

where R is the mean radius of curvature of the surface. Decomposing the stress tensor into pressure and viscous components,

$$\left(-p^1 + \tau_{nn}'^1\right) - \left(-p^2 + \tau_{nn}'^2\right) = \frac{2\sigma}{R} \tag{6.93}$$

which is the Young-Laplace equation.

(d) The thermal energy balance gives

$$q_n^1 = q_n^2 \tag{6.94}$$

That is, the heat that flows out of a medium has to flow into the other medium.

(e) The mass conservation balance of species A, at an interface without a chemical reaction, yields

$$j_{An}^1 = j_{An}^2 \tag{6.95}$$

The concentration at both sides of the interface are related by equilibrium data.

Symmetry Boundary Conditions

Symmetries are of utmost importance in engineering and physics. They allow for major simplifications of the problem and reduction of the calculation effort. We can distinguish between plane, axial and spherical symmetry. Furthermore, the effect of the symmetry is different depending on the tensor character of the fluid variable.

Plane Symmetry

Let us denote with $\boldsymbol{n}$ the normal vector to the plane of symmetry.

(a) For a scalar field φ, at the plane of symmetry, $\partial\varphi/\partial n = 0$.
(b) The velocity component orthogonal to the plane of symmetry must vanish at the plane of symmetry, $\boldsymbol{n} \cdot \boldsymbol{v} = 0$ and its normal derivative with respect to the plane of symmetry is zero, $\partial\boldsymbol{n} \cdot \boldsymbol{v}/\partial n = 0$. The velocity component parallel to the plane of symmetry behaves like a standard scalar.

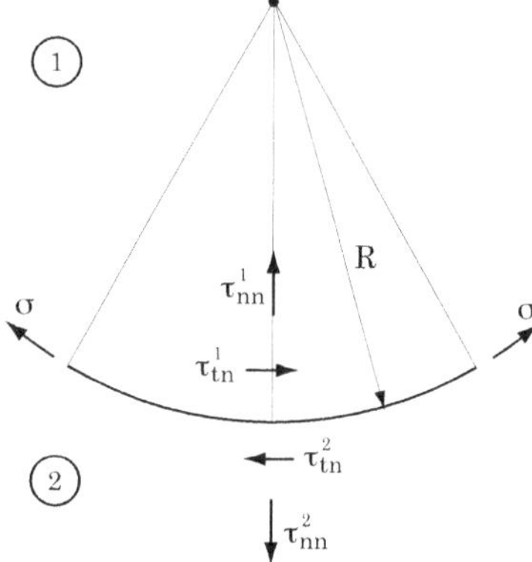

Fig. 6.9. Effect of the surface tension on the boundary conditions at interfaces.

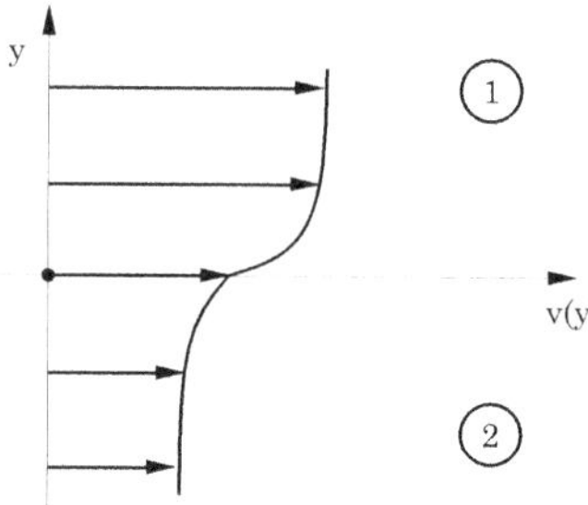

Fig. 6.10. Across an interface, the velocity and the tangential stress are continuous, but not the derivative of the velocity.

Axial Symmetry

In this case, the flow is symmetric around an axis of symmetry. This is also called symmetry of revolution. This type of flow is best analyzed using cylindrical coordinates, r, θ, z, with z the axis of symmetry (see Appendix F).

(a) For a scalar field φ, at the axis of symmetry $\partial\varphi/\partial r|_{r=0} = 0$. Furthermore, there are no variations with respect to θ, $\partial\varphi/\partial\theta = 0$.
(b) The velocity component orthogonal to the axis of symmetry must vanish at the axis of symmetry, $v_r = 0$ at $r = 0$ and $\partial v_r/\partial r|_{r=0} = 0$. The tangential velocity component cancels out everywhere, $v_\theta = 0$. The velocity component parallel to the axis of symmetry, v_z, behaves like a standard scalar, $\partial v_z/\partial r|_{r=0} = 0$.

Spherical Symmetry

In this case, the flow is symmetric around a center of symmetry. This is also called spherical symmetry. This type of flow is best analyzed using spherical coordinates, r, θ, ϕ, with $r = 0$ at the center of symmetry (see Appendix F).

(a) For a scalar field φ, at the axis of symmetry $\partial\varphi/\partial r|_{r=0} = 0$. Furthermore, there are no variations with respect to θ and ϕ, that is, $\partial\varphi/\partial\theta = 0$ and $\partial\varphi/\partial\phi = 0$.
(b) The radial velocity vanishes at the center of symmetry, $v_r = 0$ at $r = 0$. Furthermore, $\partial v_r/\partial r|_{r=0} = 0$. The tangential velocity components cancel out everywhere, $v_\theta = v_\phi = 0$.

Problems

6.1   Two plane discs of radius R are separated by a distance b. The upper plate moves downwards at a constant speed V. The fluid between both plates is squeezed so it flows in the radial direction. Determine the outflow volumetric flux Q and the maximum fluid velocity for the cases below.

(a) The outflow velocity is uniform.
(b) The outflow velocity profile is parabolic. (Consider the approximation: $v(z) = a_0 + a_1 z + a_2 z^2$ and find a_i, $0 \leq i \leq 2$, as a function of R, b and $V_{\max}$).

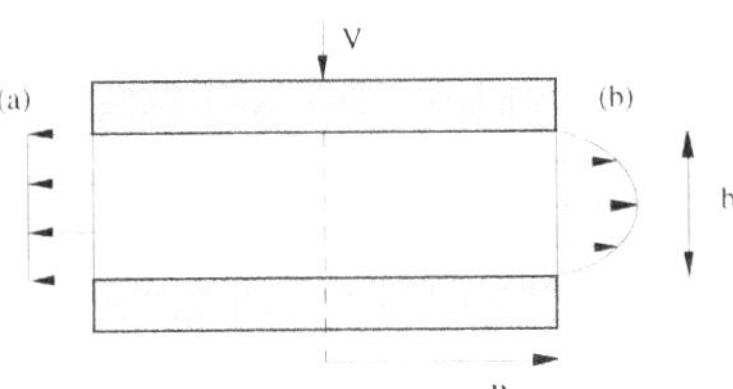

Problem 6.1.   The fluid squeezed between two parallel discs flows in the radial direction.

6.2   The syringe of the figure contains a fluid of density ρ. What is the relation between the piston speed V and the outflow volumetric flux Q ?

6.3   A tank of volume 0.05 m^3 contains air at $p = 800$ kPa absolute pressure and $T = 15\ ^\circ$C. At $t = 0$ the valve opens and air escapes at a velocity of $v = 311$ m/s and a density $\rho = 6.13\,\text{kg/m}^3$ through an opening of $A = 65$ mm^2. What is the rate of change of density $d\rho/dt$ in the tank at $t = 0$.

6.4   The reactor of the Figure has a depth of 6 m. Due to a chemical reaction in the interior, the generated gas exits through the four openings. The gas velocity and density at the exit are given by

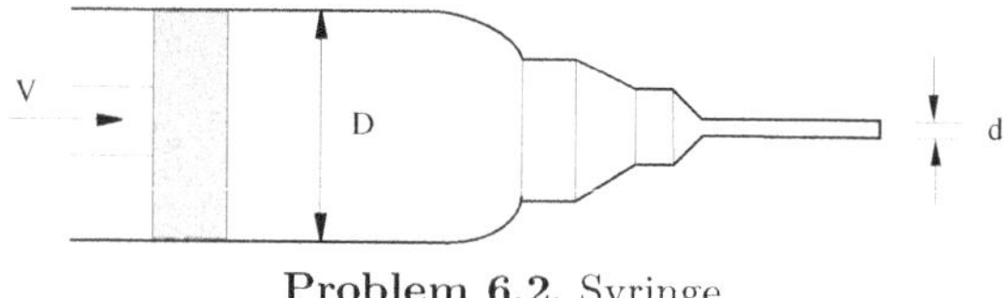

Problem 6.2. Syringe.

$$v = \frac{10}{r} \qquad \rho = 0.002 + 0.001r$$

Calculate the rate of change of the mass in the reactor.

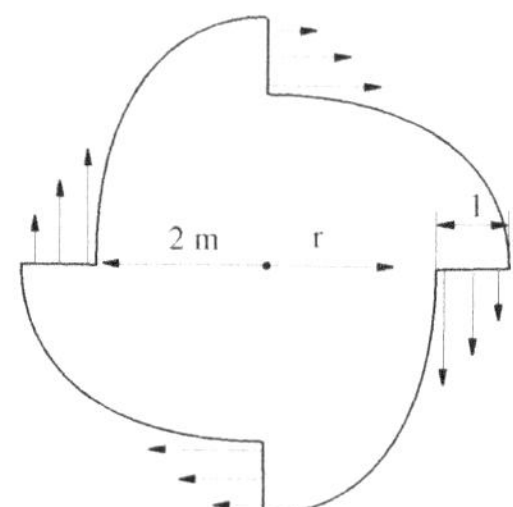

Problem 6.4. Reactor with four outlets.

6.5 Through the 180° elbow of the Figure circulates water and discharges at section 2 to the atmosphere. The gage pressure at 1 is $P_{m1} = 96$ kPa. Also, $A_1 = 2\,600$ mm^2; $A_2 = 650$ mm^2; $V_1 = 3.05$ m/s. Calculate the force F_x to hold the elbow.

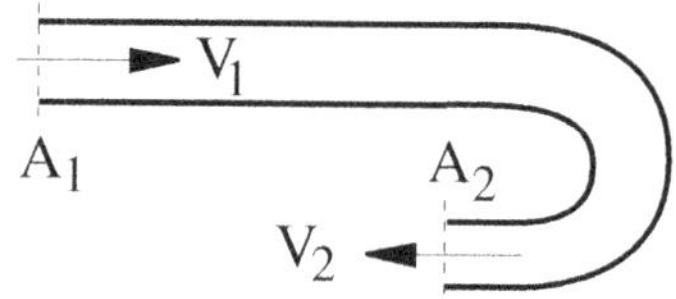

Problem 6.5. Force to hold a 180° elbow.

6.6 A perfectly equilibrated weight W is supported with a vertical jet. If the diameter of the jet is D_0, what is the jet velocity?

6.7 A bin feeds gravel into a moving belt at a rate of 65 kg/s. The diameter of the belt wheels is 80 cm and they rotate at 150 rpm. Estimate the power to operate the belt. Hint: First calculate F_x and from there, the power.

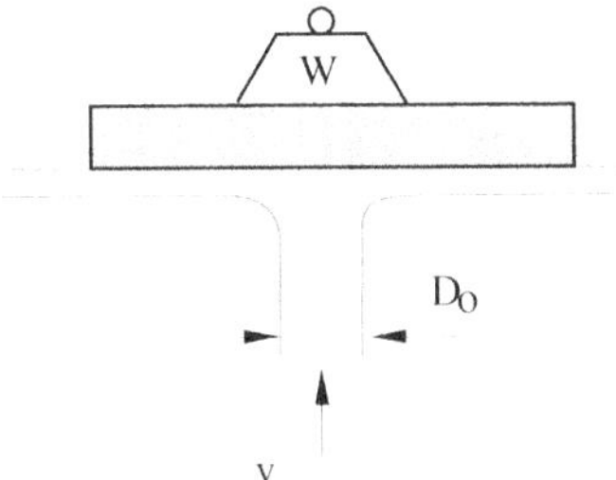

Problem 6.6. A high speed jet can hold a weight.

Problem 6.7. Power to operate a moving belt.

6.8 In the pipe of the Figure flows an incompressible fluid. Taking into account gravity, determine the force that the fluid exerts on the pipe.

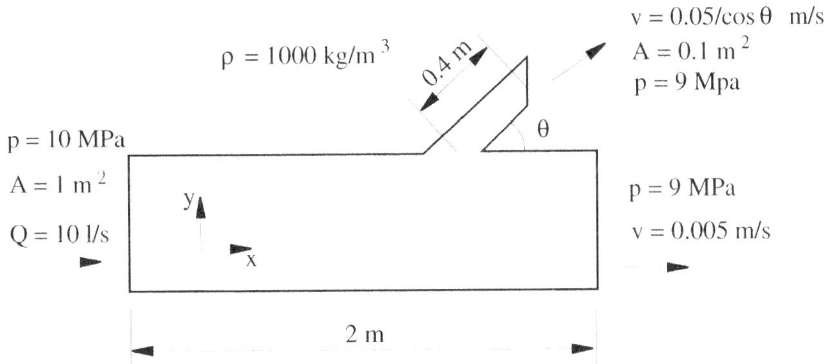

Problem 6.8. Pipe with an inclined bifurcation.

6.9 The Figure shows a cart whose top surface guides the flow and ejects it at an angle θ with respect to the horizontal line. A fixed nozzle thrusts the liquid jet at a constant speed V, propelling the cart at the constant speed V_c.

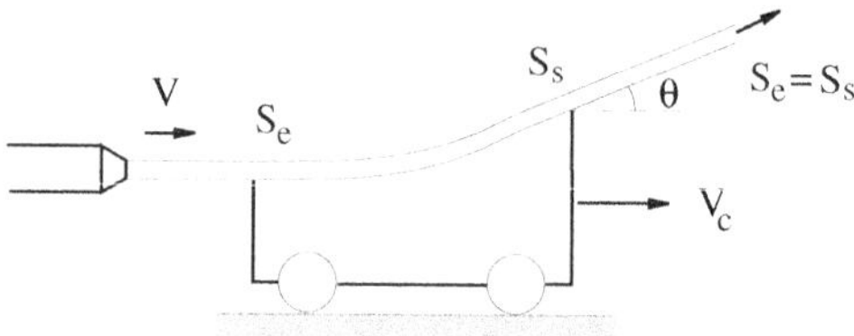

Problem 6.9. Cart propelled by a liquid jet.

(a) If the cart moves in a straight line, show that the power transferred to the cart is maximum when $V_c/V = 1/3$.
(b) For gurus: Assuming that there is a large number of guides, like that shown in the Figure, and that they are welded to a wheel whose surface turns at the linear velocity of V_c, show that the power transferred to the cart is maximum when $V_c/V = 1/2$.

6.10 The three-jet water sprinkler shown in the Figure uses the volumetric flux 2.7 m^3/h. Neglecting the friction at the turning axis and the friction with the air, what is the rotation speed for $\theta = 0°$ and $\theta = 40°$?

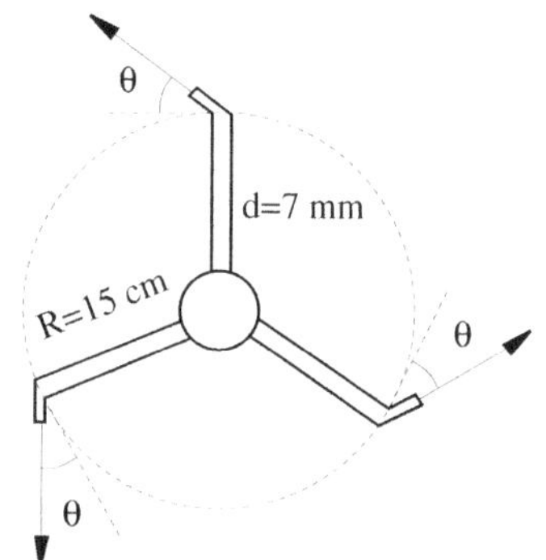

Problem 6.10. Three-jet sprinkler.

6.11 Water is forced into the device of the Figure through the inlet A at a rate of 0.1 m^3/s. At the same time, through the inlet B, oil with a relative density of 0.8 enters at a rate of 0.03 m^3/s. If the liquids are incompressible and form a homogeneous mixture of oil droplets in water, what is the mean velocity and density mixture at the outlet C, which has a diameter of 0.3 m?

6.12 In the previous problem, the piston at D has a diameter of 150 mm and moves towards the left at a velocity of 0.3 m/s. What is the mean velocity of the fluid exiting at C?

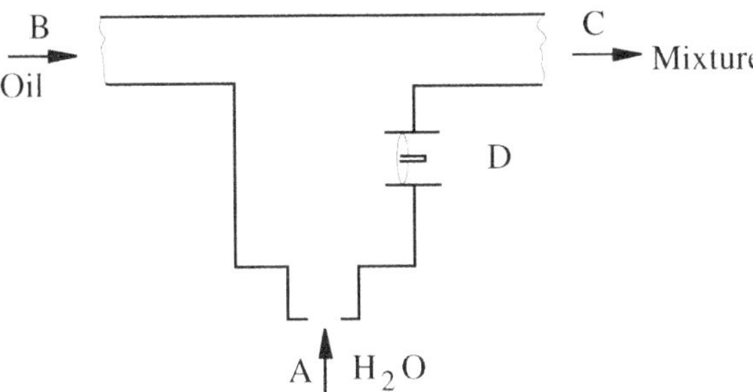

Problem 6.11. Perfectly agitated mixing tank.

6.13   The Figure depicts a 10 m wide rectangular ditch with an inclined bottom. Water flows into the ditch at a rate of $Q = 100$ l/s. What is dh/dt when $h = 1$ m? How long does it take for the level to increase from $h = 1$ m to $h = 1.2$ m?

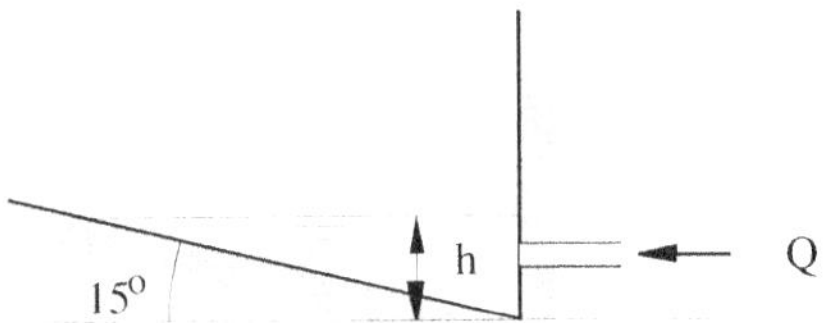

Problem 6.13. Spill into a rectangular ditch.

6.14   A liquid flows from A to B across the gradual contraction of the Figure with a volumetric flux of 54 l/s. The losses in the pipe due to friction are equal to a head of $h_f = 0.135$ m. When the pressure head at B is $h_p = 61$ cm, what is the pressure head at A? Note: $h_p = p/(\rho g)$ and $h_f = D_v/(\rho g Q)$; $D_v = \int_{V_c(t)} \phi_v \, dV$.

6.15   Let us consider the pipe network of the Figure, where the viscous dissipation between 1 and 4 is $D_v = 23$ W. Find the pressure p_3 when the inlet pressures are $p_1 = p_2 = 10$ atm and the outlet pressure, $p_4 = 1$ atm. Data: $R_1 = 2.5$ cm, $R_2 = 2.5$ cm, $R_3 = 1.25$ cm, $R_4 = 2.0$ cm, $R_5 = 2.0$ cm, $R_6 = 2.0$ cm. $u_1 = 0.01$ m/s, $u_2 = 0.1$ m/s, $u_3 = 0.05$ m/s.

6.16   At a sudden expansion, the cross-section area increases from A_1 to $A_2 > A_1$. The incompressible flow has a uniform velocity V_1 at the inlet A_1. Far enough downstream, the velocity is uniform and equal to V_2.

(a) Determine V_2 as a function of V_1, A_1 and A_2.

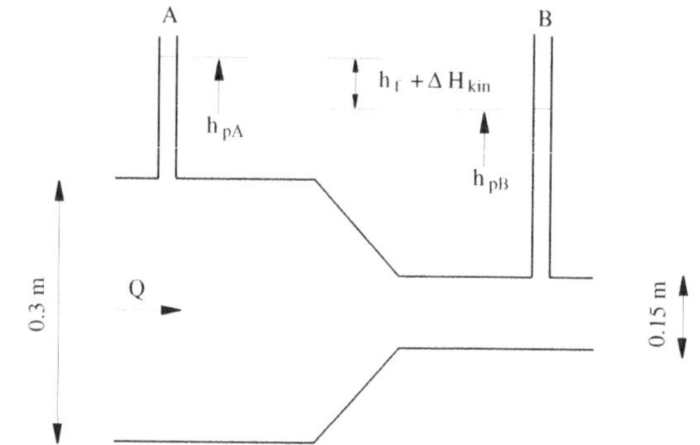

Problem 6.14. Pressure fall at a gradual contraction.

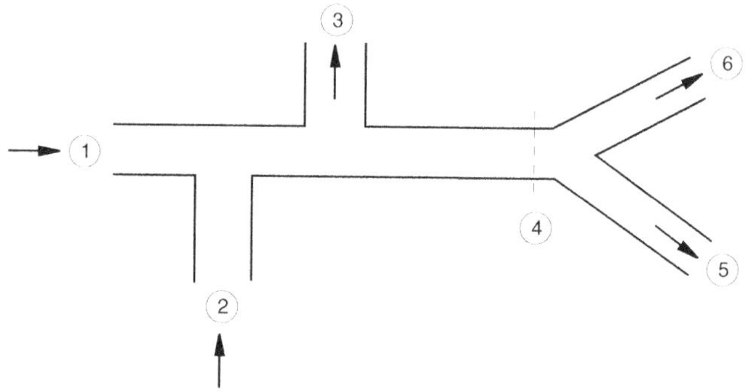

Problem 6.15. Pipe network.

(b) Determine $\Delta p/\rho$, that is, the pressure recovery across the expansion as a function of V_1, A_1 and A_2. Neglect the viscous forces at the conduit walls.

(c) Obtain the viscous dissipation D_v between the sections 1 and 2.

(d) Taking into account that the energy losses are given by the viscous dissipation D_v, determine the coefficient of local losses K_s of the expansion, using as a reference velocity V_1 (see Chapter 12).

6.17 In a mixing tank, a fluid of density ρ and specific heat c_p is stirred by an agitator that is turned at ω [rad/s] by a torque M. Calculate the heat transferred per unit time $\dot{Q}$ [W] between the tank and the environment to keep a constant temperature of 55 °C.

6.18 The conical tank of the Figure is initially full. The tank is opened and because of gravity the liquid pours. Calculate the time for the tank to empty. Tips: Relate V_2 and $\mathrm{d}h/\mathrm{d}t$ using the mass conservation equation. Then, apply the mechanical energy equation to relate V_2 as a function of $h(t)$. Combine

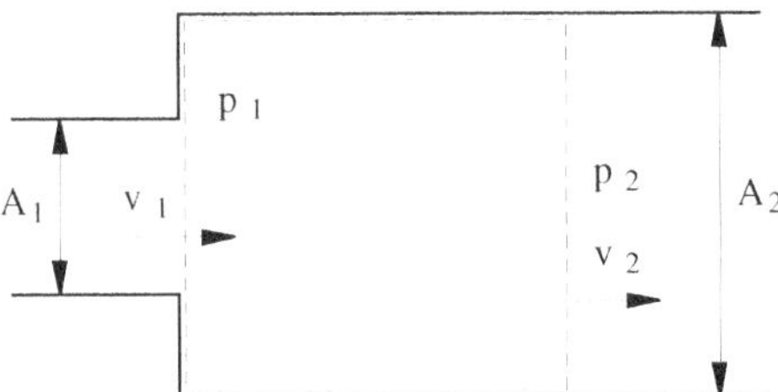

Problem 6.16. At a sudden expansion the energy losses are large.

both expressions. Neglect the time variation of kinetic energy within the tank and assume $z_2 \ll h(t)$. Initially $h(0) = z_0$.

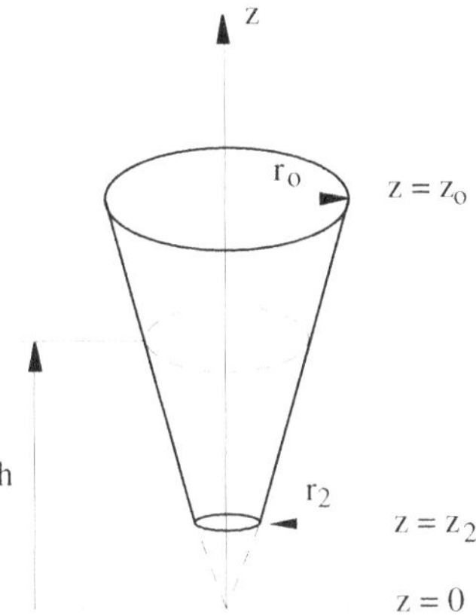

Problem 6.18. Time to empty a tank.

6.19 Consider a tank filled with a mixture of water and salt. At $t = 0$ the salt concentration is $\rho_s(t = 0) = \rho_{s0}$. Through the inlet, fresh water is introduced at a volumetric flow rate of Q. Calculate the salt concentration $\rho_s(t)$ as a function of time t, Q and the tank volume V.

6.20 In a perfectly stirred tank of 1 m^3 takes place a first-order exothermic reaction with a rate of extinction per unit volume of $0.1c$ mol/(l s) and a heat production of 5 cal/mol. The initial reactant concentration and temperature are $c_0 = 1$ mol/l and 25 °C, respectively. Calculate:

(a) The rate of consumption of the reactant in the tank.
(b) The concentration evolution of the reactant as a function of time.

(c) The heat generation per unit time due to the reaction. What happens to
the produced heat?

6.21   In the previous problem, if the reactor is adiabatic and the fluid has a
constant density ρ, a specific heat c_p and an initial temperature T_0, calculate
the temperature evolution of the fluid.

6.22   Water at 20 °C, with density $\rho = 1$ gr/cm^3, specific heat $c_p =$
4.18 J/(kg K), viscosity 1 cp and salt concentration $c_0 = 0.1$ gr/cm^3 enters a
mixer with a mean velocity $u_1 = 1$ m/s through a pipe of section $A_1 = 10$ cm^2.
Pure water at the same temperature enters the mixer through a pipe of section
$A_2 = 5$ cm^2 at the mean speed $u_2 = 0.5$ m/s. By heat addition the $V = 1$ m^3
mixer is maintained at the constant temperature of $T = 40$ °C and agitation
provides 50 W of external work. Through a pipe of section $A_3 = 7$ cm^2, the
mixture exits the tank at the temperature T and concentration c of the fluid
in the mixer.
Calculate:

(a) The exit mean velocity u_3.
(b) The salt concentration in the mixer.
(c) The heat added per unit time $\dot{Q}$.
(d) What would happen to the concentration and temperature if no heat were
added?
(e) Determine the viscous dissipation inside the mixer, D_v.

6.23   Show that the mass conservation equation for a single liquid of density
ρ reduces to the volume conservation equation.

Constitutive Equations

In the previous chapter, the equations describing transport phenomena in fluids have been introduced. However, these equations cannot be solved yet. More information about the behavior of the particular fluid is necessary. For example, the flow field will vary depending on the substance being a liquid or a gas, on behaving like a Newtonian fluid or not, and so on. Therefore, the solution of the problem will depend on equations that describe the specific behavior of the fluid, called equations of state and constitutive equations.

7.1 Introduction

In order to predict the evolution of a fluid field, we need to solve the nonlinear system of transport equations. In the general case, the equations to be solved encompass the continuity equation, the $n_{\mathrm{esp}} - 1$ chemical species equations, the momentum equation and the total energy equation.

Table 7.1. Number of equations and unknowns in the three-dimensional transport equations.

Equation	num. of equations	main unknowns	other unknowns	num. of unknowns
mass conservation	1	ρ		1
chemical species cons.[1]	$n_{\mathrm{esp}} - 1$	Y_A	$(\boldsymbol{v} - \boldsymbol{v}_A)$	$4(n_{\mathrm{esp}} - 1)$
momentum	3	v_i	p, $\boldsymbol{\tau}'$	$3 + 1 + 6$
total energy	1	e	$\boldsymbol{q}$	$1 + 3$
	$5 + n_{\mathrm{esp}} - 1$		$<$	$15 + 4(n_{\mathrm{esp}} - 1)$

For three-dimensional flows, Table 7.1 summarizes the number of equations and unknowns involved. It can be concluded that the number of unknowns exceeds the number of equations and, therefore, to solve a transport problem, information about the fluid must be supplied. Certainly, and as an example, the flow field for a gas will be different from that for a liquid. The information about the substance is given in the form of *equations of state* and *constitutive equations*.

The *equations of state* relate thermodynamic variables. For instance,

$$\left.\begin{aligned} \rho &= \rho(p,T) \\ e &= e(p,T) \end{aligned}\right\} 1 \text{ eq.}$$

The *constitutive equations* describe other fluid behavior, in particular, transport due to diffusion mechanisms, like

$$\begin{aligned} \rho_A(\boldsymbol{v}_A - \boldsymbol{v}) &= \cdots & 3(n-1) \text{ eqs.} \\ \boldsymbol{\tau}' &= \cdots & 6 \\ \boldsymbol{q} &= \cdots & 3 \end{aligned}$$

This chapter is devoted to the constitutive equations.

Remark 7.1. The equations of state and the constitutive equations characterize the behavior of the fluid and, therefore, are obtained experimentally.

7.2 Momentum Transport by Diffusion

Viscosity is responsible for this transport phenomenon. This property creates friction between the layers of fluid because of the relative motion between the fluid particles.

Example 7.1 (Agitation as a promoter of mixing). Let us consider a still mixing tank. When the agitator starts to turn, it sets in motion a layer of fluid around the blades. Due to friction, this initial fluid motion is propagated to other layers, until all the fluid in the tank is in motion. At typical Reynolds numbers, the turbulence intervenes in the process. Only for very small Reynolds numbers (very low fluid velocities) would momentum transport be due only to viscosity.

The nature of viscous diffusion transport was revealed by the Couette experiment. In this experiment, a fluid flows between two parallel infinite

---

[1] As explained in Chapter 6 only $n_{\rm csp} - 1$ chemical species equations are linearly independent.

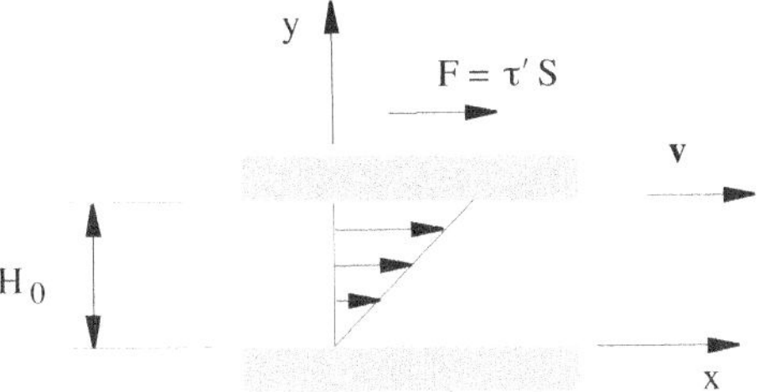

Fig. 7.1. Couette flow.

plates (see Fig. 7.1). The top plate is horizontally displaced at a speed V, while the bottom plate is kept at rest. After reaching the steady state, the force F to move the top plate is measured. From the force and the plate surface, the viscous stress τ' is calculated. Measurements for various plate distances H are plotted in Fig. 7.2 and lead to the observations below.

(a) For a given plate distance H_0, the relation between τ' and V is linear.
(b) The slope of the line τ'-V, $\tan\gamma$, is inversely proportional to the distance H,

$$\frac{\tan\gamma_0}{\tan\gamma} = \frac{H}{H_0}$$

Then, the viscous stress at the wall can be calculated as

$$\tau' = \frac{F}{S} \; = \tan\gamma \, V$$
$$= \tan\gamma_0 \tfrac{H_0}{H} \, V$$
$$= \underbrace{\tan\gamma_0 H_0}_{\mu} \, \underbrace{\frac{V}{H}}_{\frac{du}{dy}}$$

Note that $\frac{du}{dy}$ (where u is the horizontal velocity component) equals V/H for a linear velocity profile and generalizes the experimental observations to nonlinear velocity profiles.

Definition 7.1 (Dynamic viscosity). *The proportionality constant between the viscous stress τ' and the velocity gradient is called dynamic viscosity μ. Its dimensions are $[\mu] = \mathsf{ML^{-1}T^{-1}}$ and its units in the SI, Pa s or kg/(m s).*

Newton's law of friction

The above findings are expressed by the *Newton's law* of friction

$$\boxed{\tau' = \tau'_{xy} = \mu \frac{du}{dy}} \tag{7.1}$$

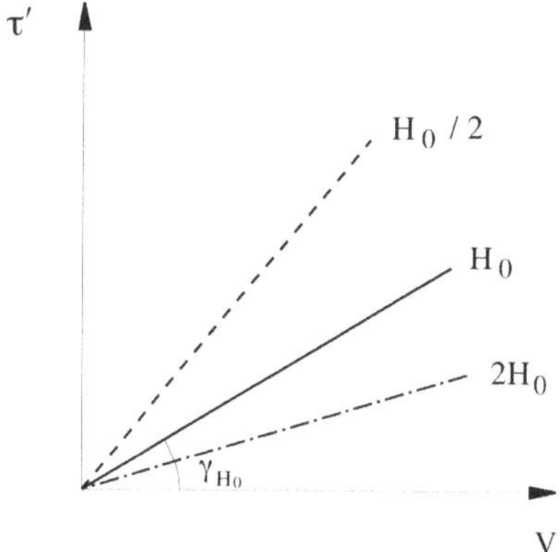

Fig. 7.2. Lines τ'-V as a function of the plates distance.

Law of Navier-Poisson

The generalization of Newton's law of friction to three-dimensional flows is known as the *law of Navier-Poisson*, which in indicial notation and Cartesian coordinates is

$$\tau'_{ij} = \mu \left(\frac{\partial v_i}{\partial x_j} + \frac{\partial v_j}{\partial x_i} \right) + \lambda\, v_{k,k}\delta_{ij} \tag{7.2}$$

where μ is the dynamic viscosity and λ, the *second viscosity coefficient*. Expanding (7.2)

$$\boldsymbol{\tau}' = \mu \begin{pmatrix} 2\dfrac{\partial v_1}{\partial x_1} & \dfrac{\partial v_1}{\partial x_2} + \dfrac{\partial v_2}{\partial x_1} & \dfrac{\partial v_1}{\partial x_3} + \dfrac{\partial v_3}{\partial x_1} \\[2mm] & 2\dfrac{\partial v_2}{\partial x_2} & \dfrac{\partial v_2}{\partial x_3} + \dfrac{\partial v_3}{\partial x_2} \\[2mm] \text{symm.} & & 2\dfrac{\partial v_3}{\partial x_3} \end{pmatrix} + \lambda v_{k,k} \begin{pmatrix} 1 & 0 & 0 \\ 0 & 1 & 0 \\ 0 & 0 & 1 \end{pmatrix}$$

Definition 7.2 (Bulk viscosity). *The coefficient $B = \lambda + \frac{2}{3}\mu$ is called bulk viscosity, and it is related to friction and dissipation associated with spherically symmetric compression/expansion processes.*

Remark 7.2. According to the *Stokes hypothesis*, it is usually assumed that

$$\lambda = -\frac{2}{3}\mu$$

This value implies that spherically symmetric expansion and contraction processes are frictionless, i.e., $B = 0$.

Remark 7.3. The second law of thermodynamics dictates that $\mu \geq 0$ and $\lambda + 2/3\mu \geq 0$.

Definition 7.3 (Kinematic viscosity). *The kinematic viscosity ν is the ratio*

$$\boxed{\nu = \frac{\mu}{\rho}} \tag{7.3}$$

Its dimensions are $[\nu] = \mathsf{L}^2/\mathsf{T}$.

Remark 7.4. In the CGS system, the units of dynamic and kinematic viscosity are, respectively, the poise, $1\ \mathrm{P} = 1\ \mathrm{g/(cm\ s)} = 0.1\ \mathrm{kg/(m\,s)}$, and the stokes, $1\ \mathrm{St} = 1\ \mathrm{cm^2/s} = 10^{-4}\ \mathrm{m^2/s}$. Very often, the centi-poise (cP) and the centi-stokes (cSt) are used. For reference, it is useful to know that the viscosity of water at $20\ {}^\circ\mathrm{C}$ is approximately 1 cP and 1 cSt.

Remark 7.5. Tables 7.2 and 7.3 show the viscosity of several substances. Note that the viscosity is a temperature dependent function. For gases, the viscosity tends to increase with temperature, whereas for liquids, to decrease.

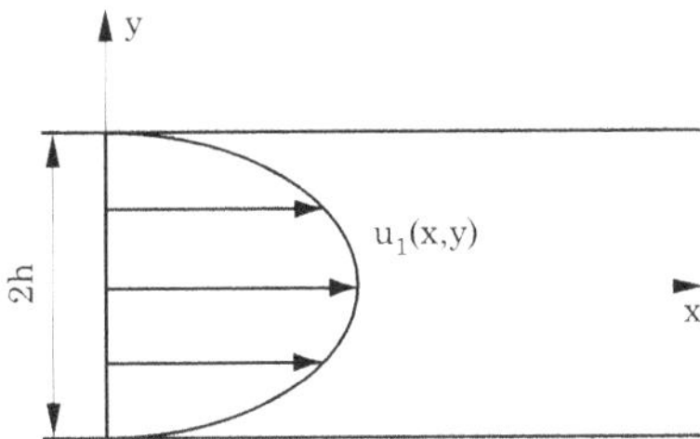

Fig. 7.3. Plane Hagen-Poiseuille flow.

Example 7.2 (Viscous stress in Poiseuille flow). In the steady, incompressible, plane Hagen-Poiseuille flow (see Fig. 7.3), determine the viscous stresses.
Solution. The steady Hagen-Poiseuille flow corresponds to the fully developed velocity field between two parallel infinite plates, separated by a distance $2h$, and is given by

$$\boldsymbol{u}(x,y) = \left\{ \begin{array}{c} u_1(x,y) \\ u_2(x,y) \end{array} \right\} = \left\{ \begin{array}{c} V_0\left[1 - \left(\frac{y}{h}\right)^2\right] \\ 0 \end{array} \right\}$$

where y is the axis perpendicular to the plates and x is aligned with the plates and the flow direction, placed at the axis of symmetry. The maximum centerline velocity V_0 depends on the pressure gradient and the viscosity.

The divergence of the velocity field can be calculated as

$$u_{k,k} = u_{1,1} + u_{2,2} = 0$$

and, therefore, the flow is incompressible. The viscous stress tensor is

$$\boldsymbol{\tau}'(x,y) = \mu \begin{bmatrix} 2u_{1,1} & u_{1,2}+u_{2,1} \\ u_{1,2}+u_{2,1} & 2u_{2,2} \end{bmatrix} = \mu \begin{bmatrix} 0 & -\frac{2V_0}{h}\frac{y}{h} \\ -\frac{2V_0}{h}\frac{y}{h} & 0 \end{bmatrix}$$

At the bottom wall $y = -h$, the wall shear stress is

$$\tau'_{12}\Big|_{y=-h} = \tau'_0 = 2\mu V_0/h$$

so the viscous stress tensor can be written as

$$\boldsymbol{\tau}'(x,y) = \begin{bmatrix} 0 & -\tau'_0\frac{y}{h} \\ -\tau'_0\frac{y}{h} & 0 \end{bmatrix}$$

An important concept that enters into the constitutive relation of the viscous stress is the rate of deformation.

Definition 7.4 (Deformation rate or strain rate). *Given a velocity field, v, the deformation or strain rate S is the symmetric part of the velocity gradient tensor,*

$$\boldsymbol{S} = \frac{1}{2}(\boldsymbol{\nabla v} + (\boldsymbol{\nabla v})^T) \tag{7.4}$$

In Cartesian components,

$$S_{ij} = \frac{1}{2}\left(\frac{\partial v_i}{\partial x_j} + \frac{\partial v_j}{\partial x_i}\right) \tag{7.5}$$

The rate of deformation represents how fast the fluid particle deforms. The deformation is expressed as change of angles and change of unitary volume per unit time of a fluid particle. Its dimensions are $[\boldsymbol{S}] = \mathsf{T}^{-1}$.

Using the strain rate, the Navier-Poisson constitutive equation (7.2) can be written as

$$\boldsymbol{\tau}' = 2\mu\,\boldsymbol{S} + \lambda\boldsymbol{\nabla}\cdot\boldsymbol{v}\,\boldsymbol{\delta}$$

Rheology

Rheology is the branch of mechanics that studies the deformation of substances and their viscosity as a function of the relevant variables.

Definition 7.5 (Rheograms). *The behavior of the viscous shear stress as a function of the deformation rate is displayed in the curves called rheograms.*

Table 7.2. Viscosity of liquid water and air at 1 atm [3].

T	Liquid Water		Air	
	$\mu \times 10^3$	$\nu \times 10^6$	$\mu \times 10^5$	$\nu \times 10^5$
°C	kg/(m s)	m^2/s	kg/(m s)	m^2/s
0	1.787	1.787	1.716	1.327
20	1.0019	1.0036	1.813	1.505
40	0.653	0.6581	1.908	1.692
60	0.4665	0.4744	1.999	1.886
80	0.3548	0.3651	2.087	2.088
100	0.2821	0.2944	2.173	2.298

Table 7.3. Viscosity of several gases at 1 atm [3].

Gas	T	$\mu \times 10^5$
	°C	kg/(m s)
CH_4	20	1.09
H_2O	100	1.211
CO_2	20	1.46
N_2	20	1.75
O_2	20	2.04

For the most frequently encountered substances, such as water and air, the viscosity depends only on thermodynamic variables, like the temperature, and is independent on the velocity gradient or deformation rate. These fluids are called *Newtonian* and the constitutive relation between the viscous stress and the deformation rate is *linear*. Other examples of Newtonian fluids are oils, glycerine, monomers, liquid metals, milk and honey.

However, this is not the case for all substances, where the viscosity may be a function of the deformation rate. In this case, the fluid is called *non-Newtonian*, and the constitutive equation is a *nonlinear* function of the deformation rate. Examples of this type of fluid are paintings, mayonnaise, plaster, slurries, etc.

Fig. 7.4 shows the rheogram of a Newtonian fluid compared with that of various classes of non-Newtonian substances: *dilatant, pseudoplastic* and *ideal plastic (Bingham)*.

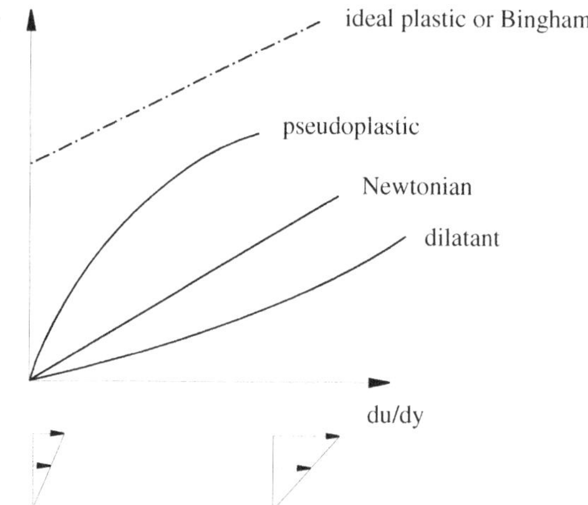

Fig. 7.4. Rheogram.

In order to clarify the behavior of non-Newtonian substances, we introduce the *apparent viscosity*,

$$\eta_a = \frac{\tau'}{du/dy}$$

which coincides with the viscosity μ for Newtonian fluids. For non-Newtonian fluids, the apparent viscosity is a function of the deformation rate.

Classification of non-Newtonian Fluids

According to the behavior of the viscous stress, non-Newtonian fluids can be classified as follows.

(a) *Dilatant or shear-thickening:* the apparent viscosity increases with the deformation rate. Dilatant fluids are typically multi-phase fluids, like fluids with bubbles or particles. For small deformation rates, the bubbles behave as if they were not present. But at large deformation rates, the bubbles or particles start to collide, increasing the friction and, as a consequence, the viscosity. Examples: corn-starch, suspensions, emulsions, dispersions and mixtures.

(b) *Pseudoplastic or shear-thinning:* the apparent viscosity decreases with the deformation rate. Pseudoplastic fluids are fluids made of large chains. At small deformation rates, the chains are disorganized and tangled, causing a large apparent viscosity. As the deformation rate increases, the chains organize, tending to align with the flow, decreasing the viscosity. Examples: solutions of polymers and melted polymers, foams, painting, food sauces.

(c) *Bingham or viscoplastic:* these substances behave like a solid until a threshold on the stress is exceeded. Then, they start to flow, either as Newtonian or non-Newtonian fluids. When the slope of the rheogram is constant, the substance is called *ideal plastic.* Examples: toothpaste, ketchup, mayonnaise, butter, paintings, mud, wax, blood.

(d) *Viscoelastic fluids:* these are fluids that at the same time present characteristics of fluids (which are viscous) and solids (which are elastic). The behavior of these fluids depends on their history, so they are said to have memory effects and can recover their original state. The constitutive relations for these fluids are very complex, even transport equations. Examples: melted polymers, polymers solutions, egg-white, dough and tar.

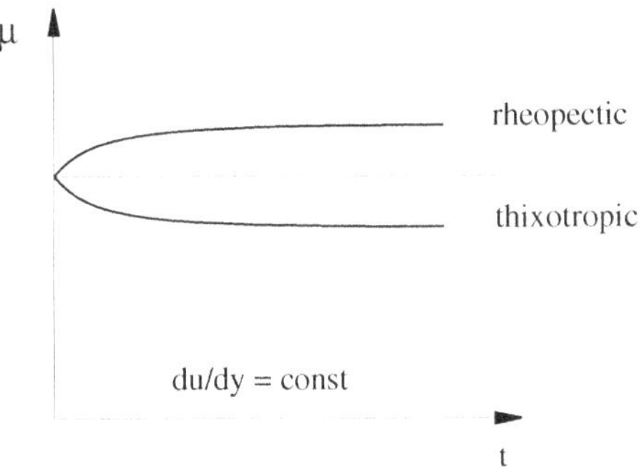

Fig. 7.5. Evolution of the viscosity on time.

Time Dependency

For certain substances, the viscosity can depend on the time that the fluid has been subject to stress (Fig. 7.5). This is an example of viscosity depending on history. When the viscosity decreases or increases with time, the fluid is said to be *rheopectic* or *thixotropic*, respectively.

In this case, when subject to cyclic rheograms, the substances exhibit *hysteresis.*

Constitutive Equations for Non-Newtonian Fluids

A very simple constitutive relation for non-Newtonian fluids is the *power law,* which is capable of representing dilatant and pseudoplastic fluids,

$$\tau' = \mu \left| \frac{du}{dy} \right|^{n-1} \frac{du}{dy} \tag{7.6}$$

Note that Newtonian fluids are recovered for $n = 1$. For other cases, the apparent viscosity can be calculated as

$$\eta_a = \mu \left| \frac{\mathrm{d}u}{\mathrm{d}y} \right|^{n-1} \tag{7.7}$$

If $n > 1$, the apparent viscosity increases with the deformation rate (dilatant fluid), whereas for $n < 1$, it decreases (pseudoplastic fluid).

The behavior of a *Bingham fluid* can be represented by the law

$$\begin{cases} \tau' \geq \tau'_0 & \tau' = \tau'_0 + \mu \left| \frac{\mathrm{d}u}{\mathrm{d}y} \right|^{n-1} \frac{\mathrm{d}u}{\mathrm{d}y} \\ \tau' < \tau'_0 & 0 = \frac{\mathrm{d}u}{\mathrm{d}y} \end{cases} \tag{7.8}$$

where τ'_0 is the *stress threshold* from which the substance starts to flow, separating the solid and liquid behavior.

Another example of constitutive relation for a non-Newtonian fluid is the *Ellis law*,

$$\tau' = \mu \left[1 + \mu \left| \frac{\mathrm{d}u}{\mathrm{d}y} \right|^n \right]^{-1} \frac{\mathrm{d}u}{\mathrm{d}y} \tag{7.9}$$

where the Newtonian case is recovered for $n = 0$.

A very extended model for viscoelastic fluids is the *Oldroyd model*, where

$$\tau' + \lambda_1 \frac{\mathrm{d}\tau'}{\mathrm{d}t} = \mu \left(\frac{\mathrm{d}u}{\mathrm{d}y} + \lambda_2 \frac{\mathrm{d}}{\mathrm{d}t} \frac{\mathrm{d}u}{\mathrm{d}y} \right)$$

The constants λ_1 and λ_2 are relaxation times and μ the viscosity.

The viscoelastic behaviour of these fluids is characterized by the dimensionless Weissenberg and Deborah numbers. These numbers are similar, but not identical. The Weissenberg number can be defined as the product of the relaxation time t_r times the strain rate (which is a shear inverse time scale),

$$\mathrm{Wi} = t_r S_{12}$$

The relaxation time t_r measures the time that the system uses to respond to an external action, lets say, of time t_0. Whereas the Deborah number is the ratio between the relaxation time t_r and a characteristic flow time scale,

$$\mathrm{De} = \frac{t_r}{t_0}$$

It characterizes how "fluid" a substance is.

Example 7.3 (Fluency of a Bingham fluid). Let the constitutive equation of a Bingham fluid of density ρ and fluency stress τ_0 be given by

$$\tau' = \tau_0 + \eta \frac{\mathrm{d}u}{\mathrm{d}y}$$

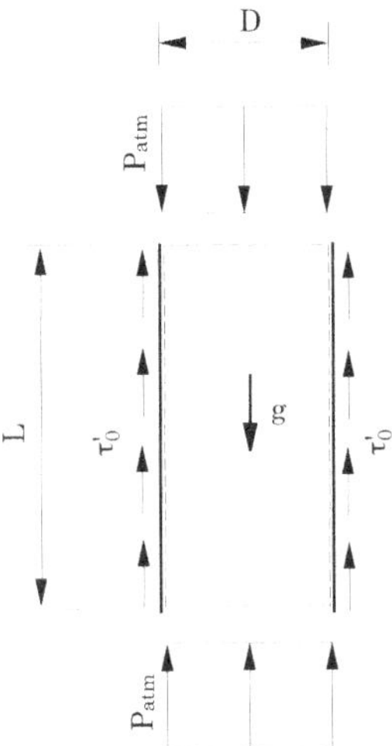

Fig. 7.6. Fluency threshold of a Bingham fluid in a vertical pipe due to gravity.

The fluid fills a vertical circular section pipe of length L. Determine the minimum radius R necessary for the substance to start flowing due to the gravity acceleration g.

Solution. For a Bingham fluid to start to flow, it is necessary that the viscous stress exceeds τ_0, the threshold stress of fluency. Assuming steady, fully developed flow, the vertical momentum equation gives the force balance at the point of fluency,

$$\tau_0 \, 2\pi R L = \rho \pi R^2 L g$$

from where

$$R = \frac{2\tau_0}{\rho g}$$

If the radius is smaller, the fluid will not flow by gravity. The fluency stress is very important when designing plants for Bingham fluids.

7.3 Heat Transport by Diffusion

Heat diffusion is the mechanism of heat transport by molecular interaction. It can take place in both solids and fluids.

In order to investigate the nature of heat transport the experiment of Fig. 7.7 is set up. Two parallel plates, separated a distance H, are heated at constant temperature T_1 and $T_2 < T_1$. In the steady state, it is observed that the temperature distribution is linear and that the heat flux q [W/m^2] transferred from the hot plate to the cold plate obeys the following laws.

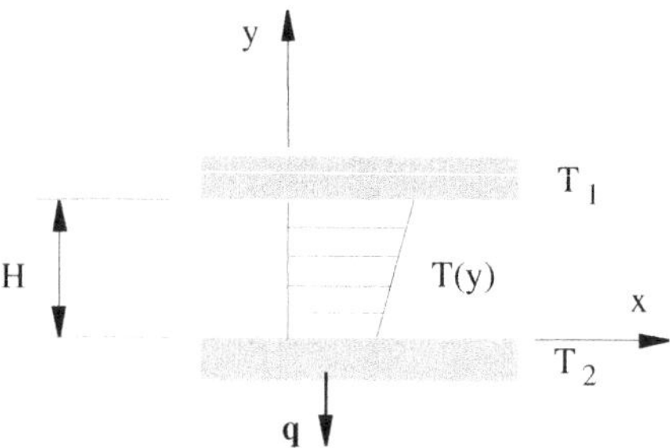

Fig. 7.7. Experiment for heat diffusion.

(a) The heat transfer is proportional to the temperature difference between the plates, $\Delta T = T_1 - T_2$.
(b) The heat transfer is inversely proportional to the distance between the plates, H.

Mathematically, this can be expressed as

$$q = \frac{\dot{Q}}{S} = \kappa \frac{\Delta T}{H}$$

where κ is the proportionality constant.

In order to generalize the previous result to nonlinear temperature distributions, the finite increments are substituted by the derivative of the temperature with respect to space,

$$q_y = -\kappa \frac{\partial T}{\partial y} \tag{7.10}$$

which is called the *Fourier's law* of heat conduction. The minus sign is added because the heat travels from areas of large temperature to areas of small temperatures, that is, in the opposite direction to the temperature gradient.

Definition 7.6 (Thermal conductivity). *The proportionality coefficient between the heat flux and the temperature gradient, κ, is called thermal conductivity and its units in the SI are $[\kappa] = \mathrm{W/(m\,K)}$.*

Fourier's law

The three-dimensional form of Fourier's law is

$$\boxed{q = -\kappa\, \nabla T} \tag{7.11}$$

Example 7.4. Above 1 km height, the temperature in the troposphere (from 0 to 11 km height) decreases approximately 6.5 °C per km. If the thermal conductivity of air is about 0.0241 W/(m K), calculate the vertical heat flux due to heat conduction in the troposphere.

Solution. Neglecting other types of heat transport, the conduction heat flux is calculated by Fourier's law,

$$q_z = -\kappa \frac{\partial T}{\partial z}$$

where z is the vertical axis. From the data of the problem,

$$\frac{\partial T}{\partial z} = -0.0065 \text{ K/m}$$

and substituting,

$$q_z = 0.0241 \times 0.0065 = 1.57 \times 10^{-4} \text{ W/m}^2$$

Remark 7.6. For nonlinear materials, the thermal conductivity is a function of the temperature $\kappa(T)$.

Remark 7.7. For non-isotropic materials, for which the heat flux does not point in the direction of the temperature gradient, the thermal conductivity is a matrix, $\boldsymbol{\kappa}$, so

$$\boldsymbol{q} = -\boldsymbol{\kappa} \nabla T$$

For isotropic materials, this matrix becomes the identity matrix times the conductivity,

$$\boldsymbol{\kappa} = \kappa \boldsymbol{I} \tag{7.12}$$

and the heat flux vector points in the direction of the temperature gradient.

Definition 7.7 (Thermal diffusivity). *The thermal diffusivity α is the ratio*

$$\boxed{\alpha = \frac{\kappa}{\rho c_p}} \tag{7.13}$$

where κ is the thermal conductivity, ρ the fluid density and c_p the specific heat at constant pressure. Note that its dimensions are L^2/T.

Remark 7.8. Tables 7.4 and 7.5 show the thermal conductivity of saturated water and various gases. The thermal conductivity of liquids and gases tends to increase with temperature, but there are exceptions.

Remark 7.9. The second law of thermodynamics implies that $\kappa \geq 0$.

Table 7.4. Thermal conductivity and diffusivity of liquid water at saturated conditions [27] and air at 1 atm [10].

	Liquid Water		Air	
T	$\kappa \times 10^3$	$\alpha \times 10^6$	$\kappa \times 10^3$	$\alpha \times 10^6$
K	W/(m K)	m^2/s	W/(m K)	m^2/s
300	608.	14.6	26.3	22.5
350	667.	16.4	30.0	29.9
400	686.	17.3	33.8	38.3
450	673.	17.1	37.3	47.2
500	635.	16.4	40.7	56.7

Table 7.5. Thermal conductivity of several gases at 1 atm [3, 10].

Gas	T	$\kappa \times 10^3$	Gas	T	$\kappa \times 10^3$
	K	W/(m K)		K	W/(m K)
CH_4	100	10.63	N_2	100	9.58
	200	21.84		200	18.3
	300	34.27		300	25.9
H_2O	400	26.1	O_2	100	9.04
	500	33.9		200	18.33
	600	42.2		300	26.57
CO_2	200	9.50			
	300	16.65			

7.4 Mass Transport by Binary Diffusion

Neglecting the effects of temperature gradients, pressure gradients and body forces on mass diffusion, for binary mixtures (mixtures of two components) the mechanics of diffusion mass transport is analogous to that of diffusion heat transport.

In order to understand this process experimentally, two parallel plates are set with constant concentrations of a substance A, ρ_{A1} and $\rho_{A2} < \rho_{A1}$. When the steady state is reached, the concentration distribution is linear. It is observed that the mass flux $j_A[\text{kg}/(\text{m}^2\ \text{s})]$ between the walls obeys the following rules.

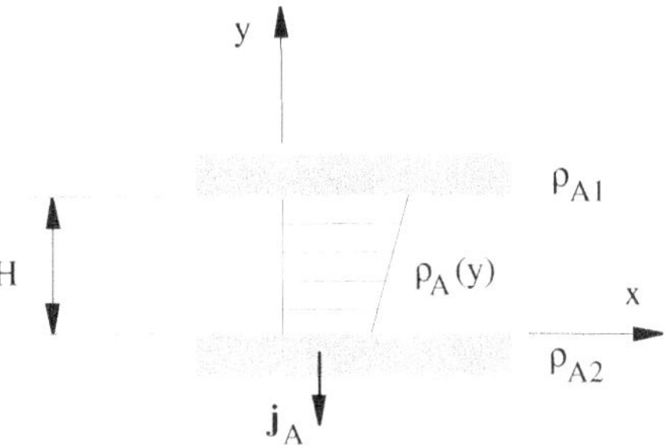

Fig. 7.8. Experiment for mass diffusion.

(a) The mass flux is proportional to the difference of concentration, $\Delta\rho_A = \rho_{A1} - \rho_{A2}$.

(b) The mass flux is inversely proportional to the distance between the plates, H.

Mathematically this can be expressed as

$$j_A = \frac{J_A}{S} = D_{AB}\frac{\Delta\rho_A}{H} \tag{7.14}$$

where D_{AB} is the proportionality constant, which represents the *diffusion of A within the fluid B*.

This is generalized as

$$j_{Ay} = -D_{AB}\frac{\partial\rho_A}{\partial y} \tag{7.15}$$

which is known as *Fick's law* of binary mass diffusion. The minus sign indicates that the mass transport is in the opposite direction from the concentration gradient, that is, from areas of large concentration to areas of small concentration.

Definition 7.8 (Molecular diffusivity). *The coefficient of proportionality, D_{AB}, is called the coefficient of mass diffusion of the substance A within the fluid B. Its units in the SI are $[D_{AB}] = m^2/s$.*

The three-dimensional expression of Fick's law for *dilute* binary mixtures is

$$\boldsymbol{j}_A = -D_{AB}\,\nabla\rho_A \tag{7.16}$$

This is a good approximation when the density of the mixture is practically constant. More general forms of Fick's law are presented below.

Fick's law for binary diffusion

Fluxes with respect to the mass average velocity

For *mass fluxes* [kg/(m^2 s)] with respect to the *mass average velocity* $\boldsymbol{v}$, Fick's law of binary diffusion states that

$$\boxed{\boldsymbol{j}_A = -\rho D_{AB}\,\nabla Y_A} \tag{7.17}$$

and for *molar fluxes* [mol/(m^2 s)]

$$\boxed{\boldsymbol{j}'_A = -\frac{\rho D_{AB}}{M_A}\,\nabla Y_A} \tag{7.18}$$

Fluxes with respect to the molar average velocity

If the mean fluid velocity is calculated using the molar average, $\boldsymbol{v}^m$, the corresponding mass fluxes by diffusion can be defined. For example, the *molar flux*, [mol/(m^2 s)], can be calculated by the corresponding Fick's law,

$$\boxed{\boldsymbol{j}_A^{m'} = -c D_{AB}\,\nabla X_A} \tag{7.19}$$

which can be particularized for constant mixture molar concentration as

$$\boldsymbol{j}_A^{m'} = -D_{AB}\,\nabla c_A \tag{7.20}$$

The above molar flux, as a *mass flux*, can be calculated as

$$\boxed{\boldsymbol{j}_A^m = -c M_A D_{AB}\,\nabla X_A} \tag{7.21}$$

Remark 7.10. The mass diffusion coefficients are symmetric and for a binary mixture $D_{BA} = D_{AB}$.

Remark 7.11. The mass diffusion coefficients typically depend on the temperature, the chemical composition and the pressure. For gases, a good approximation is

$$D = D_0 \left(\frac{T}{T_0}\right)^m \frac{p_0}{p}$$

which reflects that the diffusion coefficient D increases with temperature but decreases with pressure and the molar mass. See Table 7.6. For liquids, the diffusion coefficient increases with temperature.

Table 7.6. Coefficients for the mass diffusivity of binary mixtures of gases at $p_0 = 101\,300$ Pa and $T_0 = 273$ K [8].

Gases		$D_0 \times 10^4$	m
A	B	m^2/s	
CO	O_2	0.185	1.75
O_2	N_2	0.181	1.75
O_2	air	0.178	1.75
H_2O	air	0.22	1.75
CO_2	air	0.138	2
H_2	air	0.611	2
CH_4	air	0.196	2

Example 7.5. In a lake there is a linear concentration of salt and sediments $\rho_{\text{sed}} = \rho_0 + sz$, with z the vertical axis. Determine the mass flux of deposition of sediments at the bottom of the lake.
Solution. A mass balance at the bottom of the lake, yields that the deposition flux at the bottom equals the diffusion flux. Thus, per unit area,

$$j_{\text{sed}\,n} = -D\,\frac{\partial \rho_{\text{sed}}}{\partial n}\bigg|_{\text{bottom}} = D\,\frac{\partial \rho_{\text{sed}}}{\partial z}\bigg|_{\text{bottom}} = Ds$$

where D is the diffusivity of the sediments.

Table 7.7. Mass diffusivity of the liquid mixture ethanol/water at 25 °C [3].

A	B	T	Y_A	$D_{AB} \times 10^9$
		°C		m^2/s
Ethanol	Water	25	0.026	1.076
			0.266	0.368
			0.408	0.405
			0.680	0.743
			0.880	1.047
			0.944	1.181

Remark 7.12. The above laws of Fourier and Fick are simplified versions of the complete constitutive equations. Thermal and mass diffusion can be caused by more physical phenomena. For instance, concentration gradients can induce heat transport and, vice versa, temperature gradients can cause mass transport. See Appendix I for further details.

Table 7.8. Summary of parameters and units entering the constitutive equations.

Property	Flux	Coefficient	Diffusivity
Momentum	τ' [Pa]	Viscosity μ [kg/(m s)]	Kinematic viscosity $\nu = \frac{\mu}{\rho}$ [m^2/s]
Heat	q [W/m^2]	Thermal conductivity κ [W/(m K)]	Thermal diffusivity $\alpha = \frac{\kappa}{\rho c_p}$ [m^2/s]
Mass	j_A [kg/(m^2 s)]	Mass diffusivity D_{AB} [m^2/s]	Mass diffusivity D_{AB} [m^2/s]
Moles	j'_A [mol/(m^2 s)]	Mass diffusivity D_{AB} [m^2/s]	Mass diffusivity D_{AB} [m^2/s]

7.5 Transport Phenomena by Diffusion

The transport phenomena presented in this chapter are transport phenomena by *molecular diffusion*. In previous chapters, transport phenomena by convection have been introduced, and these have been shown to be caused by the macroscopic fluid velocity.

In contrast, the transport by molecular diffusion is caused by the random motion of the fluid molecules. This phenomenon tends to *make uniform* all the fluid properties, so the fluid evolves from local to global thermodynamic equilibrium conditions.

All the mechanisms of molecular diffusion are similarly modeled, that is, by the product of a diffusivity constant (μ, κ, D_{AB}) times a gradient of a fluid variable (see Tables 7.8 and 7.8). This is not by chance, but rather because all the diffusion mechanisms have the same origin: the microscopic random motion and collisions of the molecules. In particular:

(a) Transport of momentum. Due to the translational random motion, the molecules of smaller velocity travel to areas of higher velocity, and vice versa, causing collisions which average their velocity.

(b) Transport of heat. The temperature of a substance is an indication of its vibrational energy. Collisions between molecules at different temperature tend to average the vibrational energy of all the molecules, so the temperature becomes more uniform.

(c) Transport of mass. The molecular collisions causes the migration of molecules from areas of high concentration to areas of smaller concentrations.

Table 7.9. Elementary one-dimensional constitutive equations.

Property	Law	Equation
Momentum	Newton's law	$\tau'_{xy} = \mu \dfrac{\mathrm{d}v_x}{\mathrm{d}y}$
Heat	Fourier's law	$q_y = -\kappa \dfrac{\mathrm{d}T}{\mathrm{d}y}$
Mass	Fick's law	$j_{Ay} = -\rho D_{AB} \dfrac{\mathrm{d}Y_A}{\mathrm{d}y}$
Moles	Fick's law	$j^{m'}_{Ay} = -c D_{AB} \dfrac{\mathrm{d}X_A}{\mathrm{d}y}$

Remark 7.13. All the diffusivity coefficients ν, α, D_{AB} have the same dimensions, L^2/T.

7.6 Molecular Interpretation of Diffusion Transport

The diffusivity coefficients can be estimated for gases at low densities by the *kinetic theory of gases*. In this theory, the gas molecules are considered rigid spheres of mass m_p, which neither repel nor attract.

Let us assume that the density of molecules is N and that the molecules transport the property Γ. Γ represents the velocity $\boldsymbol{u}$ for momentum transport, specific enthalpy $c_p T$ for heat transport and chemical concentration ρ_A for mass transport of the species A.

According to the lattice model for gases, at equilibrium the relative velocity of the molecules with respect to the fluid velocity, $v_a = \sqrt{\dfrac{8KT}{\pi m_p}}$, is stochastically distributed in all spatial directions. In the above equation, K is the Boltzmann constant, T the absolute temperature and m_p the mass of each molecule. Also, spatial variations of v_a are neglected.

Let us take a surface at x with normal $\boldsymbol{n} = (1, 0, 0)$ of 1 m^2. We want to calculate the *net* flux of property Γ to the right across the surface. Thus,

$$\begin{aligned}
\Psi = \text{net flux} \quad &= \Psi_{\text{left}} - \Psi_{\text{right}} \\
&= \tfrac{1}{6} N m_p v_a \Gamma(\text{left}) - \tfrac{1}{6} N m_p v_a \Gamma(\text{right}) \\
&= \tfrac{1}{6} N m_p v_a \left(\Gamma(\text{left}) - \Gamma(\text{right}) \right) \\
&= \tfrac{1}{6} N m_p v_a \left(\Gamma(x - l/2) - \Gamma(x + l/2) \right)
\end{aligned} \tag{7.22}$$

The coefficient $1/6$ stems from distributing the molecules randomly in all three Cartesian axes and, for each axis, in the two directions. Applying Taylor series around x,

$$\begin{aligned}
\Gamma(x - l/2) \;&=\; \Gamma(x) - \frac{\mathrm{d}\Gamma(x)}{\mathrm{d}x}\frac{l}{2} + \frac{\mathrm{d}^2\Gamma(x)}{\mathrm{d}x^2}\frac{l^2}{4} - \cdots \\
\Gamma(x + l/2) \;&=\; \Gamma(x) + \frac{\mathrm{d}\Gamma(x)}{\mathrm{d}x}\frac{l}{2} + \frac{\mathrm{d}^2\Gamma(x)}{\mathrm{d}x^2}\frac{l^2}{4} + \cdots
\end{aligned} \tag{7.23}$$

and subtracting,

$$\Gamma(x - l/2) - \Gamma(x + l/2) \approx -\frac{\mathrm{d}\Gamma(x)}{\mathrm{d}x} l \tag{7.24}$$

The length l is related to the distance that the molecules travel between collisions in direction x and is proportional to the mean-free path $\lambda = 1/(\sqrt{2}\pi N d^2)$, with d the molecule diameter. Thus,

$$\begin{aligned}
\Psi \;&=\; -\frac{1}{6} N m_p v_a l \frac{\mathrm{d}\Gamma(x)}{\mathrm{d}x} \\
&= -D \frac{\mathrm{d}\Gamma(x)}{\mathrm{d}x}
\end{aligned} \tag{7.25}$$

where D is the diffusion coefficient,

$$\boxed{D = \frac{1}{6} N m_p v_a l} \tag{7.26}$$

Particularizing for each transport phenomenon,

Ψ	Γ	Diffusion Coefficient
τ'	u	$\mu = D$
q	$c_p T$	$\kappa = c_p D$
j_A	ρ_A	$D_{AA} = D$

Therefore, for a gas at low pressures, the following relation between the diffusion transport coefficients can be derived,

$$\mu \approx \frac{\kappa}{c_p} \approx D_{AA} \tag{7.27}$$

where D_{AA} is the self-diffusion parameter.

Remark 7.14. There is also a kinetic theory for liquids but, in general, it is much more technical.

Problems

7.1   A rod of 15 cm diameter turns at 1800 rpm in the interior of an orifice of 15.05 cm diameter and 30 cm length. The space between the rod and the orifice is filled with an oil of viscosity $\mu = 0.018$ kg/(m s). What is the power P necessary to equilibrate the viscous resistance to turn the rod?

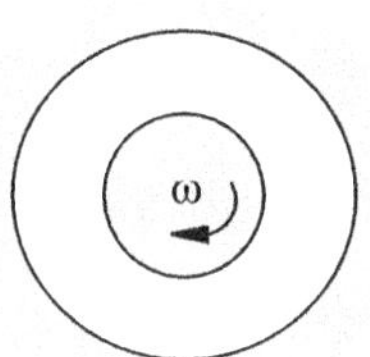

Problem 7.1. Axle turning into a concentric bearing.

7.2   If the angular velocity of the axle in the above problem is doubled, how many times does the power P increases?

7.3   A plastic panel of surface 1 m² and thickness 1 cm transports heat at a rate of 3 W at steady state when the top and bottom surface temperatures are, respectively, $T_0 = 24\,°C$ and $T_1 = 26\,°C$. What is the thermal conductivity κ of the plastic?

7.4   The space between two parallel plates separated 1.5 cm is filled with an oil of viscosity $\mu = 0.05$ kg/(m s). Between the plates, a rectangular thin flat plate with dimensions 30×60 cm is placed at 0.5 cm from the top plate. What is the necessary force to move the middle plate at 0.4 m/s?

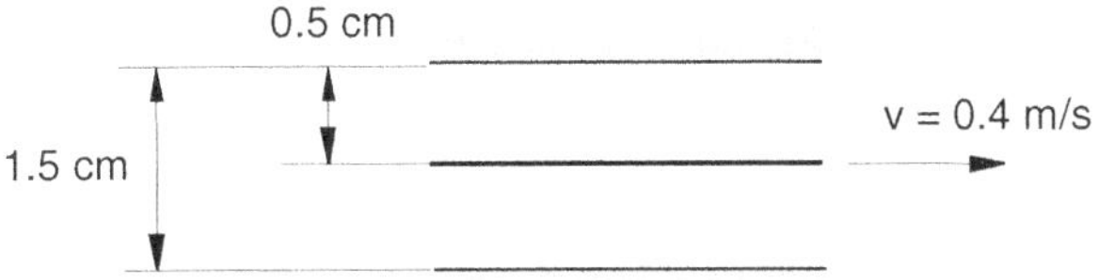

Problem 7.4.   A plate immersed in a fluid presents resistance to the motion.

7.5 On a windy day, the wind blows at 30 m/s causing the free surface of a puddle to move at a speed of 1 m/s. The thickness of the puddle is 5 cm. Assuming that the velocity profile is linear within the puddle and the air boundary layer, estimate the air boundary layer thickness δ. Data: $\mu_{air} = 1.82 \times 10^{-5}$ kg/(m s); $\mu_{water} = 1.00 \times 10^{-3}$ kg/(m s).

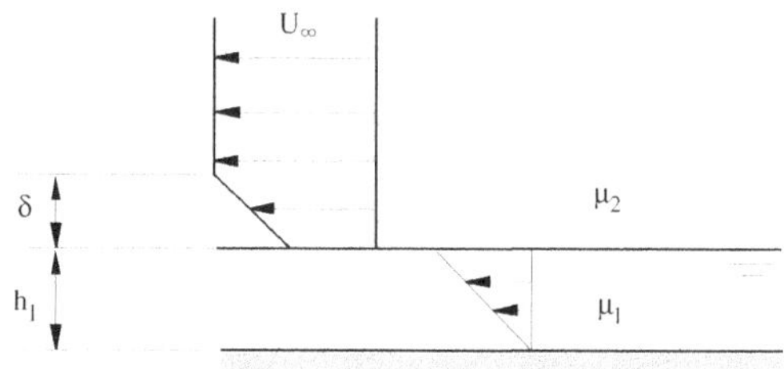

Problem 7.5. The wind drives the surface of the water by friction.

7.6 A block of mass m slides down an inclined plane as the Figure shows. Determine the terminal speed of the block (i.e. the maximum speed that the block reaches at steady state) if between the block and the plane there exists a thin film of thickness h of a fluid with viscosity μ. Assume that the velocity profile is linear in the film and that the contact area is A.

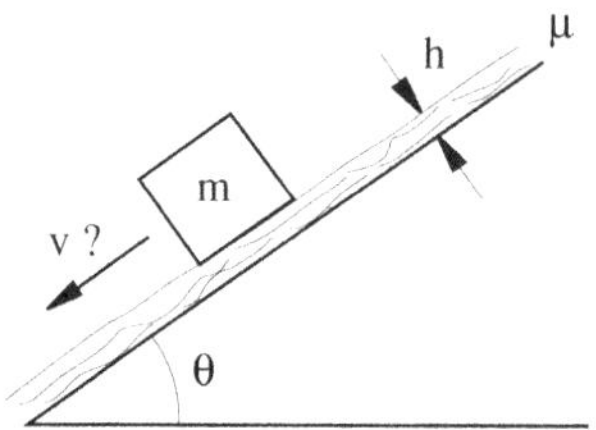

Problem 7.6. A thin film of fluid acts as a lubricant.

8

Differential Conservation Principles

As shown in Chapter 6, the integral equations provide global information about the domain of interest, such as global balances or average values of the fluid variables. Sometimes, however, more detailed information about the fluid field is required, like local temperature distributions or concentration gradients. This kind of information can be provided by the differential form of the transport equations.

The equations derived in this chapter stem from those of Chapter 6 and therefore apply for multicomponent systems with the same body force for all chemical species. For the transport equations where the body force depends on the species, see Appendix I.

8.1 Derivation of the Differential Conservation Equations

Mainly there are two ways to derive the differential transport equations from their integral counterparts.

(a) Write the conservation equations for a fixed, arbitrary control volume; transform all the integrals into volume integrals and invoke the fundamental lemma of the calculus of variations.

(b) Write the conservation equations for a fixed infinitesimal elemental control volume and then use Taylor series expansions to relate the variables at opposite sides.

In this chapter, because of its simplicity, the first procedure will be applied.

Let $V_c(t)$ be an arbitrary control volume, fixed in space, i.e $\boldsymbol{v}^c = \boldsymbol{0}$. For a general property per unit mass ϕ, the integral transport equation can be written in the form

$$\frac{\mathrm{d}}{\mathrm{d}t} \int_{V_c(t)} \rho\phi \, \mathrm{d}V + \int_{S_c(t)} \rho\phi \left[(\boldsymbol{v} - \boldsymbol{v}^c) \cdot \boldsymbol{n} \right] \mathrm{d}S = \int_{S_c(t)} \boldsymbol{t}_\phi \cdot \boldsymbol{n} \, \mathrm{d}V + \int_{V_c(t)} f_\phi \, \mathrm{d}V \tag{8.1}$$

Applying the transport theorem to the left-hand side,

$$\int_{V_c(t)} \frac{\partial \rho \phi}{\partial t}\, dV + \int_{S_c(t)} \rho \phi\, \boldsymbol{v} \cdot \boldsymbol{n}\, dS = \int_{S_c(t)} \boldsymbol{t}_\phi \cdot \boldsymbol{n}\, dV + \int_{V_c(t)} f_\phi\, dV \quad (8.2)$$

and using the Gauss divergence theorem (see Appendix E) to transform the surface integrals into volume integrals,

$$\int_{V_c(t)} \frac{\partial \rho \phi}{\partial t}\, dV + \int_{V_c(t)} \nabla \cdot (\rho \phi\, \boldsymbol{v})\, dV = \int_{V_c(t)} \nabla \cdot \boldsymbol{t}_\phi\, dV + \int_{V_c(t)} f_\phi\, dV \quad (8.3)$$

Gathering all the contributions into a single volume integral,

$$\int_{V_c(t)} \left[\frac{\partial \rho \phi}{\partial t} + \nabla \cdot (\rho \phi\, \boldsymbol{v}) - \nabla \cdot \boldsymbol{t}_\phi - f_\phi \right] dV = 0 \quad (8.4)$$

and because the domain of integration $V_c(t)$ is arbitrary, according to the fundamental lemma of the calculus of variations, the integrand must vanish:

$$\frac{\partial \rho \phi}{\partial t} + \nabla \cdot (\rho \phi\, \boldsymbol{v}) - \nabla \cdot \boldsymbol{t}_\phi - f_\phi = 0 \quad (8.5)$$

This procedure can be carried out into every integral equation to attain the results summarized below.

8.2 Continuity Equation

The integral equation of mass conservation can be recovered from the general equation (8.1) by taking $\phi = 1$, $\boldsymbol{t}_\phi = \boldsymbol{0}$ and $f_\phi = 0$. Applying the above procedure, the differential equation of mass conservation, called the *continuity equation*, is obtained

$$\frac{\partial \rho}{\partial t} + \nabla \cdot (\rho\, \boldsymbol{v}) = 0 \quad (8.6)$$

Using indicial notation, and for Cartesian coordinates,

$$\boxed{\frac{\partial \rho}{\partial t} + (\rho\, v_i)_{,i} = 0} \quad (8.7)$$

Noting that

$$\frac{D\rho}{Dt} = \frac{\partial \rho}{\partial t} + \boldsymbol{v} \cdot \nabla \rho \quad (8.8)$$

the continuity equation can also be written as

$$\boxed{\frac{1}{\rho} \frac{D\rho}{Dt} = -\nabla \cdot \boldsymbol{v}} \quad (8.9)$$

8.2.1 Particular case: incompressible fluid

For an *incompressible fluid* or a *liquid,* the density is constant and the continuity equation simplifies to

$$\boxed{\nabla \cdot \boldsymbol{v} = 0}$$
(8.10)

In indicial notation and for Cartesian coordinates it can be written in the more compact form

$$v_{i,i} = 0$$
(8.11)

Recall that the divergence of the velocity is the unit dilatation per unit time of an infinitesimal volume of fluid Vol (i.e. the rate of change of the volume of a fluid particle per unit volume),

$$\operatorname{div} \boldsymbol{v} = \nabla \cdot \boldsymbol{v} = \frac{1}{\mathrm{Vol}} \frac{\mathrm{dVol}}{\mathrm{d}t}$$
(8.12)

For a liquid, since the volume of a fluid particle (or a small fluid volume) is constant, the dilatation is zero. Thus the divergence of the velocity in an incompressible velocity field is zero.

8.3 Momentum Equation

Repeating the process shown above to the integral momentum equations yields

$$\frac{\partial \rho \boldsymbol{v}}{\partial t} + \nabla \cdot (\rho \boldsymbol{v}\boldsymbol{v}) = -\nabla p + \nabla \cdot \boldsymbol{\tau}' + \rho \boldsymbol{f}_m$$
(8.13)

In indicial notation and Cartesian coordinates,

$$\boxed{\frac{\partial \rho v_i}{\partial t} + (\rho v_i v_j)_{,j} = -p_{,i} + \tau'_{ij,j} + \rho f_{mi}}$$
(8.14)

8.3.1 Particular case: Newtonian liquid with constant viscosity

For a Newtonian incompressible fluid, since the divergence of the velocity vanishes, i.e.

$$v_{i,i} = 0$$

the Navier-Poisson constitutive equation in Cartesian coordinates simplifies to

$$\begin{aligned}
\tau'_{ij} &= \mu \left(\frac{\partial v_i}{\partial x_j} + \frac{\partial v_j}{\partial x_i} \right) + \lambda \, v_{k,k} \, \delta_{ij} \\
&= \mu \left(\frac{\partial v_i}{\partial x_j} + \frac{\partial v_j}{\partial x_i} \right) \\
&= \mu \left(v_{i,j} + v_{j,i} \right)
\end{aligned}$$
(8.15)

Taking the divergence, that is, taking the spatial derivative with respect to j,

$$
\begin{aligned}
\tau'_{ij,j} &= \left(\mu\left(v_{i,j}+v_{j,i}\right)\right)_{,j} \\
&= \mu\left(v_{i,j}+v_{j,i}\right)_{,j} &&(\mu \text{ constant}) \\
&= \mu\left(v_{i,jj}+v_{j,ij}\right) \\
&= \mu\left(v_{i,jj}+v_{j,ji}\right) &&(\text{continuous second derivative}) \\
&= \mu\, v_{i,jj} &&(\text{incompressible fluid})
\end{aligned}
\tag{8.16}
$$

Thus, the momentum equation for a Newtonian incompressible fluid with constant viscosity becomes

$$
\frac{\partial \rho \boldsymbol{v}}{\partial t} + \nabla \cdot (\rho \boldsymbol{v}\boldsymbol{v}) = -\nabla p + \mu \Delta \boldsymbol{v} + \rho \boldsymbol{f}_m
\tag{8.17}
$$

and using indicial notation

$$
\boxed{\ \frac{\partial \rho v_i}{\partial t} + (\rho v_i v_j)_{,j} = -p_{,i} + \mu\, v_{i,jj} + \rho f_{mi}\ }
\tag{8.18}
$$

The non-conservative form of the above equation is

$$
\rho \frac{\partial v_i}{\partial t} + \rho v_j v_{i,j} = -p_{,i} + \mu\, v_{i,jj} + \rho f_{mi}
\tag{8.19}
$$

8.4 Energy Equations

There exist many versions of the energy equation. All of them can be obtained as a combination of the first law of thermodynamics and other equations. Let us introduce the most widely used energy equations.

8.4.1 Total Energy Equation

Similarly to previous sections, from the total energy integral equation we can arrive at

$$
\frac{\partial \rho(e + \frac{1}{2}v^2)}{\partial t} + \nabla \cdot \left(\rho(e + \frac{1}{2}v^2)\boldsymbol{v}\right) = \nabla \cdot (\boldsymbol{\tau}\boldsymbol{v}) + \rho \boldsymbol{f}_m \cdot \boldsymbol{v} - \nabla \cdot \boldsymbol{q} + \dot{q}_v
\tag{8.20}
$$

If the body force is a function of a potential U independent of time, then

$$
\frac{\partial \rho(e + \frac{1}{2}v^2 + U)}{\partial t} + \nabla \cdot \left(\rho(e + \frac{1}{2}v^2 + U)\boldsymbol{v}\right) = \nabla \cdot (\boldsymbol{\tau}\boldsymbol{v}) - \nabla \cdot \boldsymbol{q} + \dot{q}_v
\tag{8.21}
$$

Again, in indicial notation and for Cartesian coordinates,

$$
\boxed{\ \frac{\partial \rho(e + \frac{1}{2}v^2 + U)}{\partial t} + \left(\rho(e + \frac{1}{2}v^2 + U)v_j\right)_{,j} = (\tau_{ij}v_i)_{,j} - q_{i,i} + \dot{q}_v\ }
\tag{8.22}
$$

8.4.2 Mechanical Energy Equation

This equation can be obtained from the momentum equation by scalar multiplication with the velocity field,

$$\boldsymbol{v} \cdot \left(\frac{\partial \rho \boldsymbol{v}}{\partial t} + \nabla \cdot (\rho \boldsymbol{v} \boldsymbol{v}) = -\nabla p + \nabla \cdot \boldsymbol{\tau}' + \rho \boldsymbol{f}_m \right) \tag{8.23}$$

Operating

$$\frac{\partial \rho \frac{1}{2} v^2}{\partial t} + \nabla \cdot (\rho \boldsymbol{v} \frac{1}{2} v^2) = -\boldsymbol{v} \cdot \nabla p + \boldsymbol{v} \cdot \nabla \cdot \boldsymbol{\tau}' + \rho \boldsymbol{v} \cdot \boldsymbol{f}_m \tag{8.24}$$

The first term on the right-hand side can be rewritten as follows,

$$\boldsymbol{v} \cdot \nabla p = \nabla \cdot (p\boldsymbol{v}) - p\nabla \cdot \boldsymbol{v} \tag{8.25}$$

which equals the *power of the pressure forces*, $\nabla \cdot (p\boldsymbol{v})$, minus the *expansion power*,

$$p\nabla \cdot \boldsymbol{v} = p \, v_{i,i} \tag{8.26}$$

The above term is called the expansion power because it is a function of the dilatation, the unit change of volume per unit time of an (infinitesimal) fluid particle (see Eq. (8.12)).

The second term on the right-hand side becomes

$$\begin{aligned} \boldsymbol{v} \cdot \nabla \cdot \boldsymbol{\tau}' \quad &= v_i \, \tau'_{ij,j} = (v_i \tau'_{ij})_{,j} - v_{i,j} \tau'_{ij} \\ &= \nabla \cdot (\boldsymbol{\tau}' \boldsymbol{v}) - \phi_v \end{aligned} \tag{8.27}$$

The function ϕ_v is the so-called *viscous dissipation function*,

$$\boxed{ \phi_v = \nabla \boldsymbol{v} : \boldsymbol{\tau}' = v_{i,j} \, \tau'_{ij} \geq 0 } \tag{8.28}$$

a scalar which is always *positive*. Note that we are using the Einstein summation convention (see Appendix D), so repeated indices are added up. Thus, the viscous dissipation function, in Cartesian coordinates can be expanded as

$$\begin{aligned} \phi_v \quad = \quad & v_{1,1} \, \tau'_{11} + v_{1,2} \, \tau'_{12} + v_{1,3} \, \tau'_{13} \\ + \quad & v_{2,1} \, \tau'_{21} + v_{2,2} \, \tau'_{22} + v_{2,3} \, \tau'_{23} \\ + \quad & v_{3,1} \, \tau'_{31} + v_{3,2} \, \tau'_{32} + v_{3,3} \, \tau'_{33} \end{aligned} \tag{8.29}$$

The viscous dissipation represents the energy lost per unit time (i.e. power) per unit volume due to viscosity, that is, to friction.

Finally, the equation of mechanical energy can be written as

$$\frac{\partial \rho \frac{1}{2} v^2}{\partial t} + \nabla \cdot \left(\rho \frac{1}{2} v^2 \boldsymbol{v} \right) = \nabla \cdot (\boldsymbol{\tau} \boldsymbol{v}) + p\nabla \cdot \boldsymbol{v} - \phi_v + \rho \boldsymbol{f}_m \cdot \boldsymbol{v} \tag{8.30}$$

and in indicial notation,

$$\boxed{\frac{\partial \rho \frac{1}{2}v_i^2}{\partial t} + \left(\rho \frac{1}{2}v_i^2 v_j\right)_{,j} = (\tau_{ij}v_i)_{,j} + p\,v_{i,i} - \phi_v + \rho f_{mi}v_i}$$
(8.31)

If the body forces stem from a potential which does not depend on time,

$$\begin{aligned}
\rho \boldsymbol{f}_m \cdot \boldsymbol{v} &= \rho\,(-\nabla U) \cdot \boldsymbol{v} \\
&= -\rho\,\left(\frac{\partial U}{\partial t} + \boldsymbol{v} \cdot \nabla U\right) \\
&= -\left(\frac{\partial \rho U}{\partial t} + \nabla \cdot (\boldsymbol{v}\rho U)\right)
\end{aligned}$$
(8.32)

in which case, the mechanical energy equation becomes

$$\frac{\partial(\rho \frac{1}{2}v^2 + \rho U)}{\partial t} + \nabla \cdot (\rho \boldsymbol{v}\frac{1}{2}v^2 + \rho \boldsymbol{v} U) = \nabla \cdot (\boldsymbol{\tau}\boldsymbol{v}) + p\nabla \cdot \boldsymbol{v} - \phi_v$$
(8.33)

or

$$\boxed{\frac{\partial \rho(\frac{1}{2}v_i^2 + U)}{\partial t} + \left(\rho(\frac{1}{2}v_i^2 + U)\,v_j\right)_{,j} = (\tau_{ij}v_i)_{,j} + p\,v_{i,i} - \phi_v}$$
(8.34)

8.4.3 Internal Energy Equation

As for the integral equation, the differential internal energy equation can also be obtained subtracting the mechanical energy from the total energy equation. The result is

$$\frac{\partial \rho e}{\partial t} + \nabla \cdot (\rho e \boldsymbol{v}) = -p\nabla \cdot \boldsymbol{v} + \phi_v - \nabla \cdot \boldsymbol{q} + \dot{q}_v$$
(8.35)

Using indicial notation and for Cartesian coordinates,

$$\boxed{\frac{\partial \rho e}{\partial t} + (\rho e v_j)_{,j} = -p\,v_{i,i} + \phi_v - q_{i,i} + \dot{q}_v}$$
(8.36)

Particular Case: Simple Compressible Substance

For a simple compressible substance (see Appendix H), the thermodynamic state can be written as a function of two independent thermodynamic variables and the Gibbs relation dictates

$$T\mathrm{d}s = \mathrm{d}e - \frac{p}{\rho^2}\,\mathrm{d}\rho$$
(8.37)

Considering $e(p, T)$, the above differential can be written as

$$T\mathrm{d}s = c_p\mathrm{d}T - \frac{\alpha_p T}{\rho}\,\mathrm{d}p \tag{8.38}$$

where

$$\alpha_p = \frac{1}{\rho}\left(\frac{\partial \rho}{\partial T}\right)_p \tag{8.39}$$

From (8.37), using the internal energy and continuity equations,

$$\begin{aligned}
\rho T\frac{Ds}{Dt} &= \rho\frac{De}{Dt} - \frac{p}{\rho}\frac{D\rho}{Dt} \\
&= \phi_v - \boldsymbol{\nabla}\cdot\boldsymbol{q} + \dot{q}_v
\end{aligned} \tag{8.40}$$

On the other hand, from (8.37),

$$\rho T\frac{Ds}{Dt} = \rho c_p\frac{DT}{Dt} - \alpha_p T\frac{Dp}{Dt} \tag{8.41}$$

Combining the last two equations,

$$\boxed{\rho c_p\frac{DT}{Dt} = \alpha_p T\frac{Dp}{Dt} + \phi_v - \boldsymbol{\nabla}\cdot\boldsymbol{q} + \dot{q}_v} \tag{8.42}$$

Particular Case: Thermally Perfect Fluid with Constant Thermal Conductivity

For a fluid with constant thermal conductivity κ, the diffusion term can be simplified as follows. Let us take Fourier's law,

$$\begin{aligned}
-q_{i,i} &= -\left(-\kappa T_{,i}\right)_{,i} \\
&= \kappa\, T_{,ii}
\end{aligned} \tag{8.43}$$

For a thermally perfect fluid (see Appendix H), the specific internal energy e is governed by the equation of state

$$\mathrm{d}e = c_v\,\mathrm{d}T$$

where c_v is the specific heat at constant volume. Then the temporal and convective terms can be cast as

$$\begin{aligned}
\frac{\partial \rho e}{\partial t} + (\rho e v_j)_{,j} &= \rho\frac{\partial e}{\partial t} + \rho v_j e_{,j} \\
&= \rho c_v\frac{\partial T}{\partial t} + \rho c_v v_j T_{,j}
\end{aligned} \tag{8.44}$$

In this way, the internal energy equation can be transformed into an equation for the temperature, which in non-conservative form is,

$$\boxed{\rho c_v\frac{\partial T}{\partial t} + \rho c_v v_j T_{,j} = -p\,v_{i,i} + \phi_v + \kappa\,T_{,jj} + \dot{q}_v} \tag{8.45}$$

Particular Case: Incompressible Fluid with Constant Conductivity

To the conditions of the previous section, we add that the fluid be incompressible. Since in this case the divergence of the velocity vanishes, $v_{i,i} = 0$, Eq. (8.45) simplifies to

$$\rho c_v \frac{\partial T}{\partial t} + \rho c_v v_j T_{,j} = \kappa\, T_{,jj} + \phi_v + \dot{q}_v \tag{8.46}$$

Futhermore, for a liquid the specific heats at constant volume and constant pressure are equal,

$$c_v = c_p$$

and, as a consequence, the internal energy equation can be written as

$$\boxed{\rho c_p \frac{\partial T}{\partial t} + \rho c_p v_j T_{,j} = \kappa\, T_{,jj} + \phi_v + \dot{q}_v} \tag{8.47}$$

Note that this equation can be recovered from (8.42) imposing $\alpha_p = 0$.

8.4.4 Enthalpy Equation

The enthalpy equation can be derived from the internal energy equation and the equation of state

$$h = e + \frac{p}{\rho} \tag{8.48}$$

where h is the specific enthalpy. Taking differentials

$$\mathrm{d}h = \mathrm{d}e + \mathrm{d}\frac{p}{\rho} \tag{8.49}$$

and likewise, taking the substantial derivative and multiplying by the density,

$$\rho \frac{D}{Dt} h = \rho \frac{D}{Dt} e + \rho \frac{D}{Dt}\frac{p}{\rho} \tag{8.50}$$

The last substantial derivative can be expanded as follows

$$\rho \frac{D}{Dt}\frac{p}{\rho} = \rho \left[\frac{1}{\rho}\frac{Dp}{Dt} - \frac{p}{\rho^2}\frac{D\rho}{Dt} \right] \tag{8.51}$$

$$= \frac{Dp}{Dt} - \frac{p}{\rho}\frac{D\rho}{Dt} \tag{8.52}$$

$$= \frac{Dp}{Dt} + p\nabla \cdot \boldsymbol{v} \tag{8.53}$$

where in the last step the continuity equation (8.9) has been substituted.

Plugging in the last result and the internal energy equation into Eq. (8.50)

$$\rho \frac{Dh}{Dt} = \frac{Dp}{Dt} + \phi_v - \boldsymbol{\nabla} \cdot \boldsymbol{q} + \dot{q}_v \tag{8.54}$$

Again, using indicial notation and Cartesian coordinates,

$$\boxed{\frac{\partial \rho h}{\partial t} + (\rho h v_j)_{,j} = \frac{\partial p}{\partial t} + v_j p_{,j} + \phi_v - q_{i,i} + \dot{q}_v} \tag{8.55}$$

8.5 Entropy Equation

As for the enthalpy equation, the transport equation for the entropy can be derived combining transport equations and equations of state. In particular, for a simple compressible substance (where the only reversible mode of work is compression – see Appendix H) the Gibbs equation is

$$T\mathrm{d}s = \mathrm{d}e + p\,\mathrm{d}\frac{1}{\rho} \tag{8.56}$$

where s is the specific entropy. Operating, the differential equation for the entropy becomes

$$\boxed{\frac{\partial \rho s}{\partial t} + (\rho s v_j)_{,j} - \left(\frac{\kappa T_{,i}}{T}\right)_{,i} - \frac{\dot{q}_v}{T} = \frac{\phi_v}{T} + \kappa \frac{T_{,i} T_{,i}}{T^2}} \tag{8.57}$$

The right-hand side gathers all the irreversible entropy-producing terms for a simple compressible substance: the viscous dissipation due to friction and the entropy production due to heat conduction, respectively.

Remark 8.1. For mixtures and in the presence of chemical reactions, the entropy equation is much more involved and beyond the scope of this text. Again, it can be derived from the Gibbs relation, which in this case is

$$T\mathrm{d}s = \mathrm{d}e + p\,\mathrm{d}\frac{1}{\rho} - \sum_{A=1}^{n_{\mathrm{esp}}} \mu_A^{\mathrm{chem}}\,\mathrm{d}Y_A \tag{8.58}$$

where $\mu_A^{\mathrm{chem}} = h_A - Ts_A$ is the specific chemical potential of species A. Then, the entropy production includes contributions stemming from mixing and chemical reaction.

8.6 Conservation of Chemical Species

The equation of conservation of chemical species is recovered from the general transport equation (8.1) by letting $\phi = Y_A$, $\boldsymbol{t}_\phi = -\boldsymbol{j}_A$ and $f_\phi = \dot{\omega}_A$. Recall that $\boldsymbol{j}_A$ is the mass flux vector of species A due to mass diffusion,

$$\boldsymbol{j}_A = \rho_A(\boldsymbol{v}_A - \boldsymbol{v}) \qquad \text{(no sum)} \tag{8.59}$$

The procedure of the first section yields

$$\frac{\partial \rho_A}{\partial t} + \nabla \cdot (\rho_A \, \boldsymbol{v}) = -\nabla \cdot \boldsymbol{j}_A + \dot{\omega}_A \tag{8.60}$$

In indicial notation and Cartesian coordinates

$$\boxed{\frac{\partial \rho_A}{\partial t} + (\rho_A \, v_j)_{,j} = -j_{Aj,j} + \dot{\omega}_A} \tag{8.61}$$

or as a function of the mass fraction Y_A

$$\boxed{\frac{\partial \rho Y_A}{\partial t} + (\rho Y_A \, v_j)_{,j} = -j_{Aj,j} + \dot{\omega}_A} \tag{8.62}$$

In non-conservative form, by means of the continuity equation, this equation can be written as

$$\rho \frac{\partial Y_A}{\partial t} + \rho v_j Y_{A,j} = -j_{Aj,j} + \dot{\omega}_A \tag{8.63}$$

As a function of the *averaged molar velocity*, $\boldsymbol{v}^m$, the diffusion mass flux relative to $\boldsymbol{v}^m$ is

$$\boldsymbol{j}_A^m = \rho_A(\boldsymbol{v}_A - \boldsymbol{v}^m) \qquad \text{(no sum)} \tag{8.64}$$

and the equation of chemical species becomes

$$\frac{\partial \rho_A}{\partial t} + \nabla \cdot (\rho_A \, \boldsymbol{v}^m) = -\nabla \cdot \boldsymbol{j}_A^m + \dot{\omega}_A \tag{8.65}$$

Finally, dividing by the molecular weight, M_A,

$$\frac{\partial c_A}{\partial t} + \nabla \cdot (c_A \, \boldsymbol{v}^m) = -\nabla \cdot \boldsymbol{j}_A^{m\prime} + \dot{\omega}_A' \tag{8.66}$$

where $\dot{\omega}_A' = \frac{\dot{\omega}_A}{M_A}$.

8.6.1 Particular case: constant density and constant molecular diffusivity

If both the molecular diffusivity coefficient D and the mixture density ρ are constant, the diffusion mass flux of the chemical species A can be written as

$$\begin{aligned}
-j_{Aj,j} &= -\left(-\rho D \, Y_{A,j}\right)_{,j} \\
&= \left(D \, \rho_{A,j}\right)_{,j} \\
&= D \, \rho_{A,jj}
\end{aligned}$$

in which case, the transport equation simplifies to

$$\frac{\partial \rho_A}{\partial t} + (\rho_A \, v_j)_{,j} = D \, \rho_{A,jj} + \dot{\omega}_A \tag{8.67}$$

As a function of the mass fraction Y_A, the above equation can be written in non-conservative form as

$$\boxed{\frac{\partial Y_A}{\partial t} + v_j Y_{A,j} = D \, Y_{A,jj} + \dot{\omega}_A/\rho} \tag{8.68}$$

Dividing Eq. (8.67) by the molecular weight of the species A, M_A, yields the corresponding equation for the molar concentration of species A,

$$\frac{\partial c_A}{\partial t} + (c_A \, v_j)_{,j} = D \, c_{A,jj} + \frac{\dot{\omega}_A}{M_A} \tag{8.69}$$

Very dilute mixtures of fluids can be approximated fairly well with the above hypothesis and equations, since the density is practically constant.

8.7 Summary

As occurs for the integral transport equations, the differential equations present a common pattern,

$$\frac{\partial \rho \phi}{\partial t} + (\rho \phi \, v_j)_{,j} = \mathcal{D}_\phi + \mathcal{F}_\phi \tag{8.70}$$

The two terms on the left-hand side stem from the substantial derivative, sum of the temporal and convective terms,

$$\begin{aligned}
\frac{\partial \rho \phi}{\partial t} + (\rho \phi \, v_j)_{,j} &= \rho \frac{\partial \phi}{\partial t} + \rho v_j \phi_{,j} \\
&= \rho \frac{D\phi}{Dt}
\end{aligned} \tag{8.71}$$

On the right-hand side, we encounter the diffusion term $\mathcal{D}_\phi$ and the source term $\mathcal{F}_\phi$ which, depending on the equation, may or may not be present. The variable ϕ stands for the transported property per unit mass (see Table 8.1).

Similarly, $\rho \phi$ is the density of the conserved property, that is, the amount of property per unit volume

$$\rho \phi = \frac{\text{property}}{\text{volume}} \tag{8.72}$$

See Table 8.2 for a few examples.

In the differential equations, the second term on the left-hand side (the convective term) is of the form

Table 8.1. Property per unit mass ϕ to recover the differential transport equations.

ϕ	Equation
1	mass conservation
$\boldsymbol{v}$	momentum
$e + \frac{1}{2}v^2$	total energy conservation
$\frac{1}{2}v^2$	mechanical energy
e	internal energy
h	enthalpy
s	entropy
Y_A	conservation of chemical species A

Table 8.2. Property per unit volume $\rho\phi$ to recover the differential transport equations.

Equation	$\rho\phi$
mass conservation	$\dfrac{\text{mass}}{\text{volume}} = \rho$
momentum	$\dfrac{\text{mass } \boldsymbol{v}}{\text{volume}} = \rho\boldsymbol{v}$
internal energy	$\dfrac{\text{mass } e}{\text{volume}} = \rho e$
chemical species A	$\dfrac{\text{mass } A}{\text{volume}} = \rho_A = \rho Y_A$

$$(\rho\phi v_j)_{,j} \tag{8.73}$$

which is the net flux per unit volume due to convective transport in an infinitesimal fluid volume,

$$(\rho\phi\, v_j)_{,j} = \frac{\text{net flux}(\rho\phi)}{\text{volume}} = \frac{\rho\phi Q}{\text{volume}} = \frac{\text{property}}{\text{volume}} \cdot \frac{\text{volume}}{\text{time}} \cdot \frac{1}{\text{volume}}$$
$$= \frac{\text{property}}{\text{volume} \cdot \text{time}} \tag{8.74}$$

Problems

8.1   Starting from the integral equations, derive the differential counterparts.

8.2   Using the continuity equation and assuming that all the variables are continuous with continuous first derivatives, derive that

$$\frac{\partial \rho \phi}{\partial t} + (\rho \phi \, v_j)_{,j} = \rho \frac{\partial \phi}{\partial t} + \rho v_j \phi_{,j}$$

Note that the right-hand side equals

$$\rho \frac{D\phi}{Dt}$$

that is, the density times the substantial derivative of ϕ.

8.3   In a fluid stream, the evolution of the concentration of sediments in suspension S [kg/m^3] is given by the transport equation

$$\frac{\partial hS}{\partial t} + \frac{\partial uhS}{\partial x} + \frac{\partial vhS}{\partial y} = \frac{\partial}{\partial x}\left(\epsilon h \frac{\partial S}{\partial x}\right) + \frac{\partial}{\partial y}\left(\epsilon h \frac{\partial S}{\partial y}\right) + R - w_f S$$

where h is the water level; u, v Cartesian velocity components; R the re-suspension of sediments into the stream; and w_f the sedimentation coefficient. Identify the temporal term, the convective term, the diffusion term and the source terms, distinguishing between the generation and destruction terms.

8.4   Derive the fundamental equation of fluid statics from the momentum equation.

8.5 Show that in Cartesian coordinates the viscous dissipation function can be written as

$$\phi_v = \frac{1}{2}\mu(v_{i,j} + v_{j,i})^2 + 3\lambda v_{k,k}^2 \geq 0$$

and, therefore, it is always non-negative.

8.6 Prove Eq. (8.9).

Part III

Dimensional Analysis. Theory and Applications

9

Dimensional Analysis

Even in the absence of chemical reactions, transport phenomena treat very complex physical processes. When chemical reactions are present, the complexity increases in an extraordinary manner. As seen in previous chapters, the calculation of the fluid field requires solving a set of coupled nonlinear partial differential equations. Even though the arrival of the computer allows us to obtain numerical solutions previously considered impossible, however, many industrial problems still cannot be solved in detail or with exactitude. For this to be possible, we must await an increase in the size and power of computers in various orders of magnitude. Until this happens, to uncover in detail and reliability the fluid dynamics of many industrial processes, one must resort to experiments. And the essential tool for laboratory tests is dimensional analysis.

9.1 Introduction

In previous chapters, we have seen that to solve transport problems we need to find solutions to very complicated nonlinear integro-differential equations. For this reason, very few exact or analytical solutions exist. Therefore, to get trustworthy information regarding the designed plants, it is necessary to resort to experiments in the lab or on a real scale.

As an example consider the flow around a sphere immersed in a uniform flow, as presented in Fig. 9.1. If the flow is incompressible, isothermal (at a constant temperature), without chemical reactions, the resistance force of the sphere F_D is a function of four variables,

$$F_D = f(\rho, U, D, \mu)$$

where ρ and μ are the fluid density and viscosity, respectively, U the upstream fluid velocity and D the sphere diameter.

If we take 10 data for each variable, which is not very much, we would have to carry out $10 \times 10 \times 10 \times 10 = 10^4$ experiments, only to obtain very poor

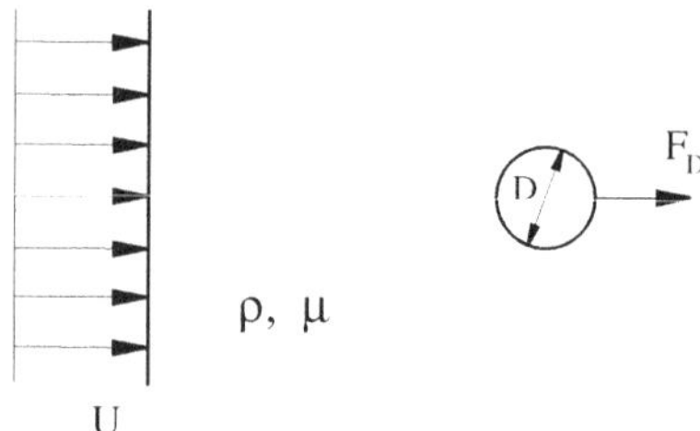

Fig. 9.1. Flow around a sphere. Problem setup.

information. As will be shown in this chapter, dimensional analysis reduces the number of experiments to be carried out to 10!

On other occasions, many systems are far too large to introduce into the laboratory. Take for example a cooling tower 100 meters high. In such situations, it is necessary to construct reduced models that fit into the lab. These models should comply with certain rules so the experimental data can be extrapolated to the real prototype. These rules of similarity are given by dimensional analysis.

Before immersing ourselves in the theory of dimensional analysis, we must recall some fundamental concepts of dimensionality.

Table 9.1. Fundamental dimensions in the SI system [5].

SI			
Dimension	Symbol	Unit	Name
Mass	M	kg	kilogram
Length	L	m	meter
Time	T	s	second
Temperature	Θ	K	kelvin
Electric current	I	A	ampere
Amount of substance	N	mol	mole
Luminous intensity	J	cd	candela

9.2 Dimensional Homogeneity Principle

All measurements are made up of magnitude, dimension and unit. The dimensions are divided into *fundamental* and *derived dimensions*, which depend upon the chosen units system. For example, in the International System (SI) and Technical System (TS), the fundamental dimensions are shown in Tables 9.1 and 9.2.

Table 9.2. Fundamental dimensions in the TS system.

TS			
Dimension	Symbol	Unit	Name
Force	F	kgf	kilogram-force
Length	L	m	meter
Time	T	s	second
Temperature	Θ	K	kelvin
Electric current	I	A	ampere

Some examples of derived dimensions in the SI are gathered in Table 9.3.

Table 9.3. Examples of derived dimensions in the SI system.

Variable	Symbol	Units	Name	Dimensions
Frequency	f	Hz	hertz	T^{-1}
Density	ρ	kg/m^3		ML^{-3}
Velocity	v	m/s		LT^{-1}
Acceleration	a	m/s^2		LT^{-2}
Force	F	N	newton	MLT^{-2}
Pressure	p	Pa	pascal	$\mathsf{ML}^{-1}\mathsf{T}^{-2}$
Energy	E	J	juole	$\mathsf{ML}^{2}\mathsf{T}^{-2}$
Power	$\dot{W}$	W	watt	$\mathsf{ML}^{2}\mathsf{T}^{-3}$

Theorem 9.1. *All equations that establish a correct physical model of behavior in a real system are dimensionally homogeneous or, likewise, are invariable when faced with a change of units.*

Example 9.1 (Uniformly accelerated motion). The distance travelled by a particle situated at x_0 at $t = 0$, with an initial speed of v_0 and then accelerated at a constant acceleration a is

$$x \;=\; x_0 \;+\; v_0 t \;+\; \tfrac{1}{2}at^2$$

$$\mathsf{L} \qquad \mathsf{L} \qquad \underbrace{\frac{\mathsf{L}}{\mathsf{T}}T}_{\mathsf{L}} \qquad \underbrace{\frac{\mathsf{L}}{\mathsf{T}^2}T^2}_{\mathsf{L}}$$

Observe that all the terms have the same dimension, L.

Definition 9.1 (Nondimensional variable). *A nondimensional variable is a variable with null dimensions, i.e. without dimensions. Sometimes it is said that the variable has dimension one.*

Example 9.2 (Nondimensional variable). In the previous example of the uniformly accelerated motion, the following nondimensional time can be defined

$$[\tau] = \left[\frac{v_0 t}{x_0}\right] = \frac{\frac{\mathsf{L}}{\mathsf{T}}\mathsf{T}}{\mathsf{L}} = 1 = [-]$$

Consequences.

(a) A combination of nondimensional variables is dimensionless.
(b) A change of units does not affect the value of a dimensionless variable.

9.3 Buckingham's Π Theorem

This theorem is the cornerstone of dimensional analysis.

Theorem 9.2. *Any dimensionally homogeneous equation which links N dimensional variables is equivalent to another equation which links $N - M$ nondimensional variables, with M being equal to or less than the number of independent dimensions of the problem.*

9.3.1 Application Process of the Π Theorem

When applying the Π theorem to solve practical cases, it is convenient to follow systematically the next steps:

1. Select the relevant variables of the problem χ_i, $i = 1, 2, \ldots, N$,

$$\varphi(\chi_1, \chi_2, \chi_3, \ldots, \chi_N) = 0$$

   One of these variables is precisely that undergoing determination. This step is one of the most difficult because it requires knowledge of the physical side of the problem: what happens, how does it happen, and so on. Therefore experience is required.

2. Construct the *dimensional matrix* and determine its range. The range is equal to M, the number of dimensionally independent variables.

Definition 9.2 (Matrix of dimensions). *If the dimensions of the variables can be expressed as*

$$[\chi_i] = \mathsf{M}^{\alpha_i} \; \mathsf{L}^{\beta_i} \; \mathsf{T}^{\gamma_i} \ldots \qquad\qquad i = 1, 2, \ldots, N$$

the matrix of dimensions is constructed with the exponents of the fundamental dimensions in the following way:

	M	L	T	$\cdots$
χ_1	α_1	β_1	γ_1	
χ_2	α_2	β_2	γ_2	
$\vdots$	$\vdots$	$\vdots$	$\vdots$	
χ_N	α_N	β_N	γ_N	

The range of the dimensional matrix M can be determined by means of Gauss elimination or by the size of the largest non-null determinant.

3. Select M dimensionally independent variables, which are to be eliminated. The following criteria should be followed.

(a) The variables to be eliminated should be dimensionally independent. This means that we can write the dimensions of the remaining variables as a function of the variables to be eliminated. For example, we cannot choose at the same time the variables $[h] = \mathsf{L}$ and $[D] = \mathsf{L}$ because both possess the same dimension, length.

(b) Simplicity.

(c) The variables that we are interested in evaluating should be maintained.

(d) Experience.

4. With the remaining $N - M$ variables, form the dimensionless groups Π_i, $i = 1, 2, \ldots, N - M$. In this way, the function φ is transformed into another function $\widetilde{\varphi}$ which depends upon dimensionless variables,

$$\varphi \longrightarrow \widetilde{\varphi}(\Pi_1, \Pi_2, \ldots, \Pi_{N-M}) = 0$$

5. Verify that the obtained dimensionless numbers Π_i do not have dimensions.

Remark 9.1. The Π theorem reduces the number of variables of a problem or physical phenomenon. The form of the function $\widetilde{\varphi}$ remains to be found by means of equations or experiments.

Example 9.3 (Resistance force of a sphere immersed in a uniform flow). Get the dimensionless variables that govern the friction of a sphere in a uniform flow.

Solution. As previously mentioned, the resistance force F_D depends upon the following variables

$$F_D = f(\rho, U, D, \mu)$$

In total we have $N = 5$ dimensional variables. We construct the dimensional matrix

	M	L	T
ρ	1	-3	0
U	0	1	-1
D	0	1	0
μ	1	-1	-1
F_D	1	1	-2

whose range is $M = 3$. The fundamental variables are chosen as ρ, U, D. With the rest of the variables (μ, F_D), $N - M = 2$ nondimensional parameters are formed as explained below.

- Π_μ requires that the following coefficient be dimensionless,

$$\Pi_\mu = \frac{\mu}{\rho^\alpha U^\beta D^\gamma}$$

Equating dimensions,

$$\mathsf{M}^1 \mathsf{L}^{-1} \mathsf{T}^{-1} = (\mathsf{ML}^{-3})^\alpha \, (\mathsf{LT}^{-1})^\beta \, \mathsf{L}^\gamma$$

which gives the linear system of equations for α, β and γ

$$\begin{cases} 1 & = \alpha \\ -1 & = -3\alpha + \beta + \gamma \\ -1 & = -\beta \end{cases}$$

and

$$\Pi_\mu = \frac{\mu}{\rho U D} \equiv \frac{1}{\mathrm{Re}}$$

The above number is the inverse of the so-called Reynolds number Re.

- Π_{F_D} requires that the following coefficient be dimensionless,

$$\Pi_{F_D} = \frac{F_D}{\rho^\alpha U^\beta D^\gamma}$$

Equating dimensions,

$$\mathsf{M}^1\mathsf{L}^1\mathsf{T}^{-2} = (\mathsf{ML}^{-3})^\alpha \, (\mathsf{LT}^{-1})^\beta \, \mathsf{L}^\gamma$$

from where

$$\begin{cases} 1 & = \alpha \\ 1 & = -3\alpha + \beta + \gamma \\ -2 & = -\beta \end{cases}$$

Note that the right-hand side is the same for all the dimensionless numbers. Finally, the result is

$$\Pi_{F_D} = \frac{F_D}{\rho U^2 D^2}$$

In practice, instead of using the above nondimensional numbers, the following numbers are used: the *Reynolds number*

$$\mathrm{Re} = \frac{\rho U D}{\mu} \tag{9.1}$$

and the *drag coefficient*

$$C_D = \frac{F_D}{\frac{1}{2}\rho U^2 A} \tag{9.2}$$

also referred to as C_f, where A is the projection of the transversal area of the object with respect to the flow and $\frac{1}{2}\rho U^2$, the kinetic energy per unit volume of the flow. Note that compared to Π_{F_D}, C_D has a dimensionless factor of $\frac{1}{2}$.

Finally, the nondimensional relationship which gives the friction of a sphere in a uniform flow simply translates into

$$\boxed{C_D = \tilde{f}(\mathrm{Re})} \tag{9.3}$$

The corresponding graphical representation can be seen in Fig. 9.2.

9.4 Applications of Dimensional Analysis

9.4.1 Simplification of Physical Equations

Going back to the example of the flow around a sphere, thanks to dimensional analysis, the drag, which initially depended upon four variables, became a function of just one variable,

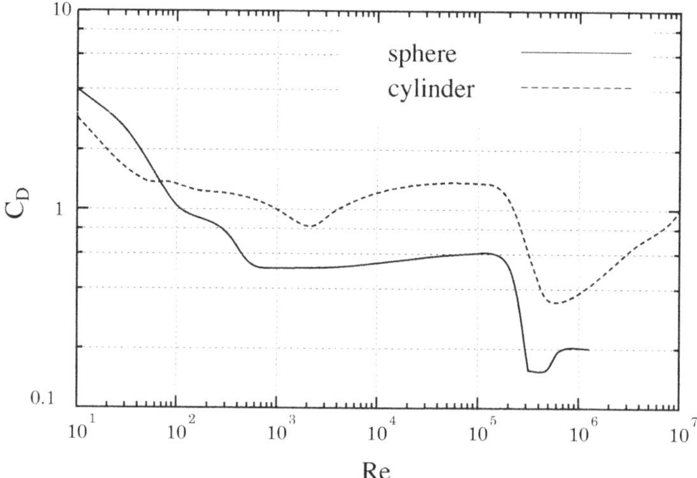

Fig. 9.2. Drag coefficient of a sphere and a circular cylinder in a uniform flow. The Reynolds number is based on the diameter D and the drag coefficient, on A, the projection of the transversal area.

$$F_D = f(\underbrace{\rho, U, D, \mu}_{4 \text{ variables}}) \quad \longrightarrow \quad C_D = \tilde{f}(\underbrace{Re}_{1 \text{ variable}})$$

This way, with just one graph as a function of the Reynolds number Re (see Fig. 9.2), we have the resistance coefficient for any diameter of the sphere, for any liquid and for any incident free-stream velocity. That's great, isn't it?

9.4.2 Experimental Economy

Imagine that we had to determine in an aerodynamic tunnel the resistance force of a sphere immersed in a uniform flow. If we began the experiment without any knowledge of dimensional analysis, and supposing that for each variable we wished to carry out ten experiments, we would need,

$$F_D = f(\underbrace{\rho, U, D, \mu}_{10 \cdot 10 \cdot 10 \cdot 10})$$

that is, 10^4 experiments.

If we use dimensional analysis and dimensionless variables, instead of carrying out 10^4 tests, we would just need to carry out 10,

$$C_D = f(\underbrace{Re}_{10})$$

This is an important saving.

Therefore, obtaining the relevant variables of a physical phenomenon with dimensional analysis allows us to reduce the number of experiments needed to be carried out.

9.4.3 Experimentation with Scaled Models. Similarity

On many occasions, the prototype to be tested does not fit into the laboratory. This is the case for very large chemical reactors, mixers, boats, airplanes, etc. In these situations, a smaller model must be made to scale, with which it is feasible to work in the laboratory.

As intuition dictates, the model must comply with a series of rules so that the results obtained in the laboratory can be extrapolated to the prototype. These rules are given by dimensional analysis and they require the existence of *similarity* between the model and the prototype.

Various kinds of similarity exist. The following are the most important. It should be noted that each type in the list is a pre-requisite to obtain the next one, and therefore it is more difficult to attain.

(a) *Geometric similarity.* This is the easiest type of similarity. The model and prototype are geometrically similar when, although being of different size, they present the same shape. This type of similarity implies that the angles of the model and the prototype are the same and that all respective lengths between the model and the prototype are in proportion.
(b) *Kinematic similarity.* This type of similarity occurs when the flow velocity in the model and prototype are related by a scale factor.
(c) *Dynamic similarity.* This type of similarity occurs when the forces and accelerations of the model and prototype are related by a scale factor.
(d) *Thermal similarity.* When the temperature differences with respect to a reference temperature in model and prototype follow a scaled reasoning.
(e) *Chemical similarity.* When the concentrations between the model and prototype follow a scaled reasoning.

The most important type of similarity is however, *complete similarity*.

Definition 9.3 (Complete similarity). *For a problem which depends upon n nondimensional variables*

$$\widetilde{\varphi}(\Pi_1, \Pi_2, \ldots, \Pi_n) = 0$$

complete similarity is said to exist between model (m) and prototype (p) if $n-1$ nondimensional variables satisfy

$$\Pi_i|_m = \Pi_i|_p$$

If for example, this occurs for $i = 1, 2, \ldots, n-1$, then

$$\Pi_n|_m = \Pi_n|_p$$

complies.

From a physical point of view, complete similarity implies that similarity exists within all the vital mechanisms of the process.

Remark 9.2. To be able to extrapolate the results from the model to the prototype, complete similarity must exist.

In many situations, it will not be possible to maintain complete similarity between the model and prototype. It may be possible to maintain only some of the dimensionless parameters between the model and prototype. In this case, it is said that *partial similarity* exists.

Definition 9.4 (Partial similarity). *When complete similarity does not exist between the model and the prototype, then partial similarity is said to exist.*

Example 9.4 (Head losses in a pipe). It is desired to circulate 600 l/s of water through a horizontal 90 cm diameter 100 m long pipe. To determine the loss of energy, an experiment is carried out in the same pipe, but using air.

(a) Under what conditions should the experiment be carried out?
(b) If the energy loss with air is $\Delta p^* = 152.88$ Pa, what power would be required to circulate the desired amount of water?

Solution. The losses in a constant cross-section straight pipe of diameter D are equal to the variation of the *modified* pressure $p^* = p + \rho g z$, with z being the height with respect to a reference level. For incompressible flows, the losses are a function of

$$\Delta p^* = f(D, L, \epsilon, U, \rho, \mu)$$

where L is the pipe length, and ρ and μ have the usual meaning. The average fluid velocity is $U = Q/S$ and the superficial finish is characterized by the roughness length ϵ [L]. Thus, there are a total of $N = 7$ dimensional variables. To be able to apply the similarity theory, as a first step the nondimensional numbers of the process must be found. The dimensional matrix is

	M	L	T
ρ	1	-3	0
U	0	1	-1
D	0	1	0
L	0	1	0
ϵ	0	1	0
μ	1	-1	-1
Δp^*	1	1	-2

whose range is $M = 3$. The chosen fundamental variables are ρ, U, D, which together with the rest of the variables form $N - M = 4$ nondimensional groups. The result is

$$\Pi_L = \frac{L}{D}$$

$$\Pi_\epsilon = \frac{\epsilon}{D}$$

$$\Pi_\mu = \frac{\mu}{\rho U D} = \mathrm{Re}^{-1}$$

$$\Pi_{\Delta p*} = \frac{\Delta p^*}{\frac{1}{2}\rho U^2}$$

The last number has been modified by the nondimensional constant $\frac{1}{2}$. Therefore,

$$\frac{\Delta p^*}{\frac{1}{2}\rho U^2} = \phi\left(\frac{L}{D}, \frac{\epsilon}{D}, \mathrm{Re}\right)$$

Experimentally it can be seen that the energy losses increase linearly with the pipe length, implying that the dependence upon L/D is linear,

$$\boxed{\frac{\Delta p^*}{\frac{1}{2}\rho U^2} = \frac{L}{D}\,\lambda\left(\frac{\epsilon}{D}, \mathrm{Re}\right)}$$

This equation is usually modified and presented in the form of the *Darcy-Weisbach equation* for the losses in head of fluid $h_f\,[\mathrm{m}]$,

$$\boxed{h_f = \frac{\Delta p^*}{\rho g} = \lambda\left(\frac{\epsilon}{D}, \mathrm{Re}\right)\frac{L}{D}\frac{U^2}{2g}}$$

The function $\lambda(\epsilon/D, \mathrm{Re})$ is determined empirically and its values are collected in correlations or in the *Moody diagram* (see Table 12.2 and Fig. 12.4).

(a) To determine how we should carry out the experiment, we shall impose similarity conditions between the model and the prototype.

Prototype	Model
Water:	Air:
$\rho_p = 1000 \mathrm{\ kg/m^3}$	$\rho_m = 1.2 \mathrm{\ kg/m^3}$
$\mu_p = 1.0 \times 10^{-3} \mathrm{\ kg/(m\,s)}$	$\mu_m = 1.81 \times 10^{-5} \mathrm{\ kg/(m\,s)}$
$L_p = 100$ m	$L_m = 100$ m
$D_p = 0.9$ m	$D_m = 0.9$ m
$Q_p = 600$ l/s	$\Delta p_m^* = 152.88$ Pa
	$Q_m = ?$

So that complete similarity may exist,

$$\left.\frac{L}{D}\right|_p = \left.\frac{L}{D}\right|_m \quad \text{(pi.1)}$$

$$\left.\frac{\epsilon}{D}\right|_p = \left.\frac{\epsilon}{D}\right|_m \quad \text{(pi.2)}$$

$$\left.\mathrm{Re}\right|_p = \left.\mathrm{Re}\right|_m \quad \text{(pi.3)}$$

which would imply that

$$\left.\frac{\Delta p^*}{\frac{1}{2}\rho U^2}\right|_p = \left.\frac{\Delta p^*}{\frac{1}{2}\rho U^2}\right|_m \quad \text{(pi.4)}$$

Given that the model test is carried out in the same pipe, $D_m = D_p$, $L_m = L_p$, $\epsilon_m = \epsilon_p$, and equalities (pi.1)-(pi.2) are trivially complied with. Eq. (pi.3) implies

$$\frac{\rho_p U_p D_p}{\mu_p} = \frac{\rho_m U_m D_m}{\mu_m}$$

from which we can deduce

$$U_m = U_p \frac{\rho_p}{\rho_m} \frac{\mu_m}{\mu_p}$$

and by multiplying by the cross-section of the tube,

$$Q_m = Q_p \frac{\rho_p}{\rho_m} \frac{\mu_m}{\mu_p} = 9\,050 \ \mathrm{l/s}$$

(b) Since the pipe is horizontal, the mechanical energy equation shows that the power necessary to drive the liquid is equal to the energy lost by friction. Given that complete similarity exists, (pi.4) is complied with,

$$\left.\frac{\Delta p^*}{\frac{1}{2}\rho U^2}\right|_p = \left.\frac{\Delta p^*}{\frac{1}{2}\rho U^2}\right|_m$$

where one can conclude that

$$\Delta p_p^* = \Delta p_m^* \frac{\rho_p U_p^2}{\rho_m U_m^2} = 560 \ \mathrm{Pa}$$

and the power to drive the water would be

$$\dot{W}_p = Q_p \Delta p_p^* = 336 \ \mathrm{W}$$

whilst the power necessary to carry out the experiment is

$$\dot{W}_m = Q_m \Delta p_m^* = 1\,384 \ \mathrm{W}$$

which is greater than the power necessary to pump water. Does any solution to reduce the air power occur to you?

Problems

9.1 Determine the dimensions of force F, stress σ, power $\dot{W}$, dynamic viscosity μ and thermal conductivity κ.

9.2 The variables which control the motion of a boat are the resistance force, F, speed V, length L, density of the liquid ρ and its viscosity μ, as well as gravity acceleration g. Obtain an expression for F using dimensional analysis.

9.3 It is believed that the power P of a fan depends upon the density of the liquid ρ, the volumetric flux Q, the diameter of the propeller D and the angular speed Ω. Using dimensional analysis, determine the dependence of P with respect to the other dimensionless variables.

9.4 In fuel injection systems, a jet of liquid breaks, forming small drops of fuel. The diameter of the resulting drops, d, supposedly depends upon the density of the liquid, the viscosity, surface tension, $[\sigma] = \text{force/length}$, and likewise upon the speed of the stream V and its diameter D. How many dimensionless parameters are required to characterize the process? Find them.

9.5 A disc spins close to a fixed surface. The radius of the disc is R, and the space between the disk and the surface is filled with a liquid of viscosity μ. The distance between the disc and the surface is h and the disc spins at an angular velocity ω. Determine the functional relationship between the torque that acts upon the disc, T, and the other variables.

9.6 A triangular weir is made of a vertical plate with an opening in the shape of a "V" with an angle ϕ cut in the upper part and transversally placed in a channel. The liquid contained in the channel is retained by the plate and obliged to flow through the opening. The discharge flow Q is a function of the raising of the liquid from the vertex of the opening. Furthermore, Q depends upon the gravity and speed at which the flow nears the weir V_0. Determine the expression that will calculate Q. What would the previous expression become if the speed V_0 was not relevant to the problem?

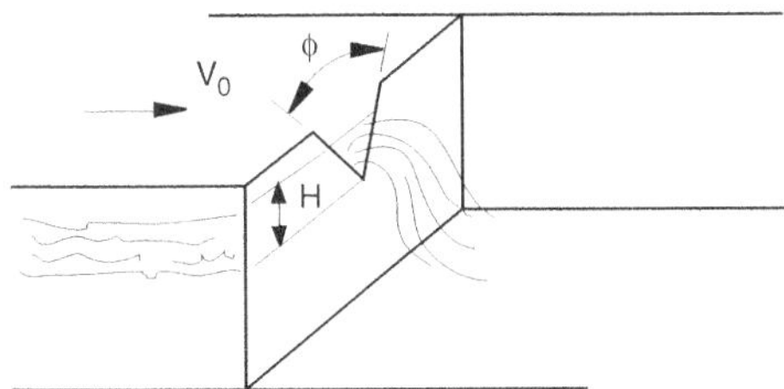

Problem 9.6. A triangular weir can be set up to measure the volumetric flux in a channel.

9.7   The drag force, F, experienced by a submarine that moves at a great depth from the surface of the water, is a function of the density ρ, viscosity μ, speed V and the transversal section of the submarine A. An expert suggests that the nondimensional relationship that allows the calculation of F is:

$$\frac{F}{\rho V^2 A} = f\left(\frac{\rho V A}{\mu}\right)$$

(a) Is the number of dimensionless parameters in the expression correct? Why?
(b) Are the parameters correct? If not, correct them.
(c) A geometrically similar model to that of the real submarine has been constructed, so that all the lengths of the model are 1/10 of those corresponding to the submarine. The model is tested in sea water.
  (1) The force of the real submarine when it moves at 5 m/s is to be determined.
  (2) At which speed should the model be tested?

9.8   An automobile must travel through standard air conditions at a speed of 100 km/h. To determine the pressure distribution, a model at a scale of 1/5 of the length of the vehicle is tested in water. Find the speed of water to be used.
$\mu_{\text{water}} = 10^{-3}$ kg/(m s), $\rho_{\text{water}} = 1\,000$ kg/m^3, $\mu_{\text{air}} = 1.8 \times 10^{-5}$ kg/(m s), $\rho_{\text{air}} = 1.2$ kg/m^3.

9.9   The depth of the steady central vortex h in a large tank of oil being stirred by a propeller needs to be predicted. One way is to carry out a study using a reduced scale model. Determine the conditions under which the experiment should be conducted to be considered a valid predictive tool. Note: Consider gh a function of gH, D, L and Ω.

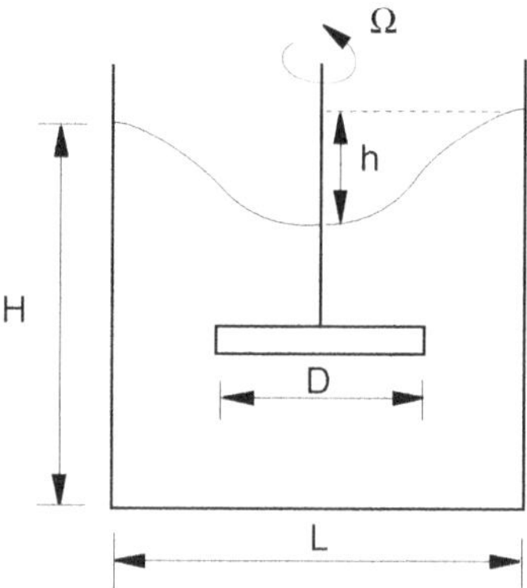

Problem 9.9. Stirring a liquid in a tank produces a vortex.

9.10   A rectangular, thin, flat plate, with dimensions of h (length) and w (width) is placed perpendicularly to a liquid current. Imagine that the drag force D which the liquid has upon the plate is a function of w and h, the density of the liquid ρ and its viscosity μ, as well as the speed V of the liquid coming towards the plate. Determine the set of dimensionless parameters to study the problem experimentally.

9.11   The Reynolds number is a very important parameter for studying transport phenomena and fluid mechanics. Estimate the Reynolds number that would be characteristic of the flow around a car traveling along the highway.

9.12   A thin layer of spherical particles are lying at the bottom of a horizontal tube, as indicated in the Figure. When an incompressible liquid flows along the tube, it can be seen that at a certain critical speed the particles move and are carried along the length of the tube. We wish to study the value of this critical speed V_c. Suppose that V_c is a function of the diameter of the tube D, the particle's diameter D_p, the liquid density ρ, the viscosity of the liquid μ, the density of the particles ρ_p and the gravity acceleration g.

(a) Using ρ, D and g as fundamental variables, obtain the dimensionless parameters of the problem.
(b) Repeat point (a) using ρ, D and μ as fundamental variables.
(c) A laboratory experiment is carried out with the same liquid and particles as the real prototype but at half the size. If a critical speed of 1 m/s is measured, what is the value of the critical speed for the real prototype in cases (a) and (b)? What is happening?
(d) Consider how this problem can be solved and calculate the critical speed in the prototype to get the critical speed of 1 m/s for the model. Which are the properties of the liquid to be used in the experiment?

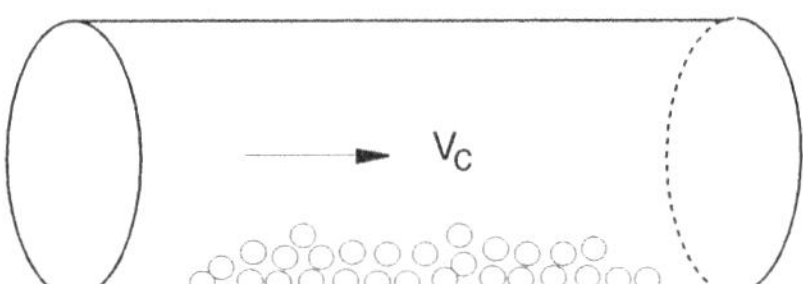

Problem 9.12.   Many industrial methods are based upon passing a liquid current over solid spheres.

9.13   During the drying process of a fine layer of liquid on a surface, the liquid evaporates and the vapor is transported in the air above the surface, as can been seen in the Figure. We are interested in knowing the dependence of the drying time t upon the rest of the variables of the problem (length L, thickness of the layer δ, the liquid's vapor pressure P_v, air speed U, viscosity μ and air density ρ).

(a) Obtain a set of dimensionless variables related to the drying time t with the rest of the variables.
(b) We wish to set up a laboratory experiment to determine the drying time of a soccer field where $P_v = 2\,000$ Pa, $L = 100$ m, $\delta = 0.01$ m and $U = 2$ m/s. In the experiment, the viscosity and the density of the air will be the same as that of the soccer field, but L will be worth 20 m (we don't have a larger laboratory available). Calculate the values of U, δ and P_v in the experiment so that complete similarity exists with the real flow.
(c) If in the experiment the average drying time is $t = 10$ min, calculate the drying time of the soccer field.

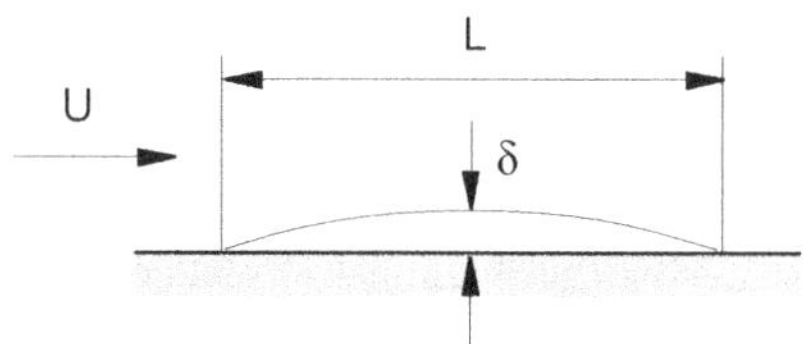

Problem 9.13. Drying process of a liquid sheet.

10

Dimensionless Equations and Numbers

The chapter on Dimensional Analysis has shown that the relevant numbers that govern the character of physical phenomena are not the dimensional variables, but rather, dimensionless numbers. In the case of partial differential equations and, in particular, the transport equations, these numbers can also be derived by making these equations dimensionless. Therefore, following this technique, this chapter extracts the fundamental dimensionless numbers that govern transport phenomena. Their physical significance will be also unveiled.

10.1 Nondimensionalization Process

In order to obtain the dimensionless numbers that characterize transport phenomena, we will use the conservation equations in differential form. The same process could be applied to the integral equations, giving the same results.

The technique to make the equations dimensionless consists of introducing a simple change of variables,

$$\boxed{\phi = \phi_0\ \phi'} \tag{10.1}$$

where ϕ is the original variable with dimensions, ϕ_0 is a dimensional constant (in particular, a reference or characteristic value of the variable) and consequently, ϕ' is a dimensionless variable.

This change of variables is introduced to all the variables and fluid properties that take part in the equations, including space and time. In particular,

$$
\begin{aligned}
t &= t_0\ t' & \boldsymbol{x} &= l_0\ \boldsymbol{x}' & \boldsymbol{v} &= v_0\ \boldsymbol{v}' \\
\rho &= \rho_0\ \rho' & \rho_A &= \rho_{A0}\ \rho'_A & p &= \Delta p_0\ p' \\
T &= \Delta T_0\ T' & \mu &= \mu_0\ \mu' & \kappa &= \kappa_0\ \kappa' \\
D_A &= D_{A0}\ D'_A & c_p &= c_{p0}\ c'_p & \boldsymbol{f}_m &= g\ \boldsymbol{f}'_m \\
\dot{\omega}_A &= \dot{\omega}_{A0}\ \dot{\omega}'_A
\end{aligned}
\tag{10.2}
$$

Remark 10.1. For a constant fluid property, it is not necessary to introduce the change of variables.

When applied to a derivative, this change of variables yields

$$\frac{\partial \phi}{\partial t} = \frac{\partial \phi_0 \phi'}{\partial t} = \phi_0 \frac{\partial \phi'}{\partial t} \tag{10.3}$$

$$\frac{\partial \phi'}{\partial t} = \frac{\partial \phi'}{\partial t'} \frac{dt'}{dt} = \frac{1}{t_0} \frac{\partial \phi'}{\partial t'} \tag{10.4}$$

Therefore,

$$\boxed{\frac{\partial \phi}{\partial t} = \frac{\phi_0}{t_0} \frac{\partial \phi'}{\partial t'}} \tag{10.5}$$

The nabla operator, being a spatial derivative, transforms in a similar way,

$$\nabla = \left\{ \begin{array}{c} \dfrac{\partial}{\partial x} \\[2mm] \dfrac{\partial}{\partial y} \\[2mm] \dfrac{\partial}{\partial z} \end{array} \right\} = \frac{1}{l_0} \left\{ \begin{array}{c} \dfrac{\partial}{\partial x'} \\[2mm] \dfrac{\partial}{\partial y'} \\[2mm] \dfrac{\partial}{\partial z'} \end{array} \right\} = \frac{1}{l_0} \nabla' \tag{10.6}$$

In conclusion

$$\boxed{\nabla \phi = \frac{\phi_0}{l_0} \nabla' \phi'} \tag{10.7}$$

10.1.1 Continuity Equation

The mass conservation equation in differential form can be written as

$$\frac{\partial \rho}{\partial t} + \nabla \cdot (\rho \, \boldsymbol{v}) = 0 \tag{10.8}$$

Introducing the above change of variables gives

$$\frac{\rho_0}{t_0} \frac{\partial \rho'}{\partial t'} + \frac{\rho_0 v_0}{l_0} \nabla' \cdot (\rho' \, \boldsymbol{v}') = 0 \tag{10.9}$$

This equation has still dimensions of ρ_0/t_0 or $\rho_0 v_0/l_0$. In order to make it dimensionless, one should divide the equation by one of the factors in front of any term.

However, in fluid dynamics and transport phenomena, by convention the factor of the *convective term* is chosen to nondimensionalize the equation. In this case, this factor is $\rho_0 v_0/l_0$, yielding

$$\frac{l_0}{v_0 t_0} \frac{\partial \rho'}{\partial t'} + \nabla' \cdot (\rho' \, \boldsymbol{v}') = 0 \tag{10.10}$$

Note that the primed terms are dimensionless and, therefore, the factor in front of the time derivative is also dimensionless. This factor is called the *Strouhal Number*

$$S = \frac{l_0}{v_0 t_0} \qquad \text{Strouhal number}$$

Finally, the dimensionless equation is

$$S\frac{\partial \rho'}{\partial t'} + \nabla' \cdot (\rho' \, v') = 0 \tag{10.11}$$

10.1.2 Momentum Equation

Starting from the differential form of the equation,

$$\frac{\partial \rho v}{\partial t} + \nabla \cdot (\rho vv) = -\nabla p + \nabla \cdot \tau' + \rho f_m \tag{10.12}$$

and applying the standard change of variables,

$$\frac{\rho_0 v_0}{t_0} \frac{\partial \rho' v'}{\partial t'} + \frac{\rho_0 v_0^2}{l_0} \nabla' \cdot (\rho' v' v') = -\frac{\Delta p_0}{l_0} \nabla' p' + \frac{\mu_0 v_0}{l_0^2} \nabla' \cdot (\tau')' + \rho_0 g \rho' f_m' \tag{10.13}$$

Dividing again by the factor pre-multiplying the convective term, $\rho_0 v_0^2 / l_0$,

$$\frac{l_0}{v_0 t_0} \frac{\partial \rho' v'}{\partial t'} + \nabla' \cdot (\rho' v' v') = -\frac{\Delta p_0}{\rho_0 v_0^2} \nabla' p' + \frac{\mu_0}{\rho_0 v_0 l_0} \nabla' \cdot (\tau')' + \frac{g l_0}{v_0^2} \rho' f_m' \tag{10.14}$$

In conclusion, the additional dimensionless numbers for the momentum equation are

$$
\begin{aligned}
\text{Eu} &= \frac{\Delta p_0}{\rho_0 v_0^2} \qquad && \text{Euler number} \\[2mm]
\text{Re} &= \frac{\rho_0 v_0 l_0}{\mu_0} \qquad && \text{Reynolds number} \\[2mm]
\text{Fr} &= \frac{v_0^2}{g l_0} \qquad && \text{Froude number}
\end{aligned}
$$

Finally, the dimensionless form of the equation is

$$S\frac{\partial \rho' v'}{\partial t'} + \nabla' \cdot (\rho' v' v') = -\text{Eu}\nabla' p' + \frac{1}{\text{Re}} \nabla' \cdot (\tau')' + \frac{1}{\text{Fr}} \rho' f_m' \tag{10.15}$$

Remark 10.2. Sometimes, one can find the Froude number as the square root of the above number,

$$\text{Fr} = \frac{v_0}{\sqrt{g l_0}}$$

Another important dimensionless number arises when the friction at a solid interface τ_0' is calculated. Indeed, the dimensionless viscous stress can be expressed as

$$(\tau_0')' = \frac{\tau_0'}{\frac{\mu_0 v_0}{l_0}} = \frac{\tau_0'}{\frac{1}{2}\rho_0 v_0^2}\frac{\frac{1}{2}\rho_0 v_0 l_0}{\mu_0} = \frac{1}{2}C_D\mathrm{Re} \tag{10.16}$$

where

$$\boxed{\quad C_D \quad = \frac{\tau_0'}{\frac{1}{2}\rho_0 v_0^2} \qquad \text{Drag coefficient}\quad}$$

10.1.3 Temperature Equation

In Chapter 8 the following temperature equation was derived for an incompressible fluid with constant heat conductivity and no volumetric heat source,

$$\frac{\rho c_p \partial T}{\partial t} + \rho c_p \boldsymbol{v} \cdot \nabla T = \kappa \nabla^2 T + \phi_v \tag{10.17}$$

The viscous dissipation function contains terms of the form

$$\phi_v = \frac{\partial u}{\partial y}\mu\frac{\partial u}{\partial y} \tag{10.18}$$

so

$$\phi_v = \frac{\mu_0 v_0^2}{l_0^2}\phi_v' \tag{10.19}$$

Nondimensionalizing the equation,

$$\frac{\rho_0 c_{p0}\Delta T_0}{t_0}\frac{\rho' c_p' \partial T''}{\partial t'} + \frac{\rho_0 c_{p0} v_0 \Delta T_0}{l_0}\rho' c_p' \boldsymbol{v}' \cdot \nabla' T'' = \frac{\kappa_0 \Delta T_0}{l_0^2}\kappa'(\nabla')^2 T'' + \frac{\mu_0 v_0^2}{l_0^2}\phi_v' \tag{10.20}$$

and dividing by the factor of the convective term,

$$\frac{l_0}{v_0 t_0}\frac{\rho' c_p' \partial T''}{\partial t'} + \rho' c_p' \boldsymbol{v}' \cdot \nabla' T'' = \frac{\kappa_0}{\rho_0 c_{p0} v_0 l_0}\kappa'(\nabla')^2 T'' + \frac{\mu_0 v_0}{\rho_0 c_{p0}\Delta T_0 l_0}\phi_v' \tag{10.21}$$

The new dimensionless numbers present are

$$\boxed{\begin{array}{ll} \mathrm{Pe} \;\; = \dfrac{\rho_0 c_{p0} v_0 l_0}{\kappa_0} & \qquad \text{Péclet number}\\[3ex] \mathrm{Ec} \;\; = \mathrm{Re}\,\dfrac{\mu_0 v_0}{\rho_0 c_{p0}\Delta T_0 l_0} = \dfrac{v_0^2}{c_{p0}\Delta T_0} & \qquad \text{Eckert number}\end{array}}$$

As a consequence, the dimensionless form of the equation is

$$S\frac{\rho' c_p' \partial T''}{\partial t'} + \rho' c_p' \boldsymbol{v}' \cdot \nabla' T'' = \frac{1}{\mathrm{Pe}}\kappa'(\nabla')^2 T'' + \frac{\mathrm{Ec}}{\mathrm{Re}}\phi_v' \tag{10.22}$$

Another important nondimensional number for heat transport phenomena arises when the diffusion heat flux at an interface is calculated. In dimensionless form, the heat flux is

$$q_0' = \frac{q_0}{\frac{\kappa_0 \Delta T}{l_0}} = \mathrm{Nu} \tag{10.23}$$

The right-hand side is the dimensionless heat flux, called

$$\boxed{\quad \mathrm{Nu} \quad = \frac{q_0 l_0}{\kappa_0 \Delta T} \qquad\qquad \text{Nusselt number} \quad}$$

10.1.4 Conservation of Chemical Species Equation

Let us take the differential form of the equation,

$$\frac{\partial \rho_A}{\partial t} + \nabla \cdot (\rho_A \, \boldsymbol{v}) = -\nabla \cdot \boldsymbol{j}_A + \dot{\omega}_A \tag{10.24}$$

Introducing the standard change of variables,

$$\frac{\rho_{A0}}{t_0} \frac{\partial \rho_A'}{\partial t'} + \frac{\rho_{A0} v_0}{l_0} \nabla' \cdot (\rho_A' \, \boldsymbol{v}') = -\frac{D_{A0} \rho_{A0}}{l_0^2} \nabla' \cdot \boldsymbol{j}_A' + \dot{\omega}_{A0} \dot{\omega}_A' \tag{10.25}$$

and dividing by the factor in front of the convective term,

$$\frac{l_0}{v_0 t_0} \frac{\partial \rho_A'}{\partial t'} + \nabla' \cdot (\rho_A' \, \boldsymbol{v}') = -\frac{D_{A0}}{v_0 l_0} \nabla' \cdot \boldsymbol{j}_A' + \frac{\dot{\omega}_{A0} l_0}{\rho_{A0} v_0} \dot{\omega}_A' \tag{10.26}$$

The new dimensionless numbers are

$$\boxed{\begin{array}{lll} \mathrm{Pe}_{\mathrm{II}} & = \dfrac{v_0 l_0}{D_{A0}} & \text{Péclet II number} \\[2ex] \mathrm{Da}_{\mathrm{I}} & = \dfrac{\dot{\omega}_{A0} l_0}{\rho_{A0} v_0} & \text{Damköhler I number} \end{array}}$$

Thus, the equation in dimensionless form can be written as

$$\mathrm{S} \frac{\partial \rho_A'}{\partial t'} + \nabla' \cdot (\rho_A' \, \boldsymbol{v}') = -\frac{1}{\mathrm{Pe}_{\mathrm{II}}} \nabla' \cdot \boldsymbol{j}_A' + \mathrm{Da}_{\mathrm{I}} \, \dot{\omega}_A' \tag{10.27}$$

Another important dimensionless number for mass transport appears when the diffusion mass flux at an interface is desired. In dimensionless form, the mass flux is

$$j_{A0}' = \frac{j_{A0}}{\frac{D_{A0} \Delta \rho_{A0}}{l_0}} = \mathrm{Nu_m} = \mathrm{Sh} \tag{10.28}$$

The right-hand side is the dimensionless heat flux, called the mass Nusselt or Sherwood number

$$\boxed{\quad \mathrm{Nu_m} = \mathrm{Sh} \quad = \frac{j_{A0} l_0}{D_{A0} \Delta \rho_{A0}} \qquad\qquad \text{Mass Nusselt or Sherwood number} \quad}$$

10.2 Other Important Dimensionless Numbers

Besides the parameters presented in the previous sections, some relevant nondimensional numbers for transport phenomena are defined as ratios of dimensionless numbers. In particular,

$$
\begin{aligned}
\mathrm{Pr} &= \frac{\mathrm{Pe}}{\mathrm{Re}} = \frac{\mu_0 c_{p0}}{\kappa_0} && \text{Prandtl number} \\[2ex]
\mathrm{Sc} &= \frac{\mathrm{Pe_{II}}}{\mathrm{Re}} = \frac{\mu_0}{\rho_0 D_{A0}} && \text{Schmidt number} \\[2ex]
\mathrm{Le} &= \frac{\mathrm{Pe_{II}}}{\mathrm{Pe}} = \frac{\kappa_0}{\rho_0 c_{p0} D_{A0}} && \text{Lewis number}
\end{aligned}
$$

Note that

$$
\mathrm{Le} = \frac{\mathrm{Sc}}{\mathrm{Pr}} \tag{10.29}
$$

Another number, significant when the fluid compressibility effects are important, is the Mach number, which relates the characteristic fluid velocity v_0 with the speed of propagation of the sonic waves in the fluid, c_0,

$$
\mathrm{Ma} = \frac{v_0}{c_0} \qquad \text{Mach number}
$$

10.3 Physical Interpretation of the Dimensionless Numbers

The dimensionless numbers are ratios that indicate the importance of a term of the equation with respect to the reference term. In particular, if the equation has been divided by the convective term, the dimensionless numbers indicate the importance of the various terms with respect to transport by convection.

These numbers can also indicate the importance of a transport phenomena compared to another. This is the case of Re, Pe, $\mathrm{Pe_{II}}$, Pr, Sc and Le.

Strouhal number

For instance, the Strouhal number represents the ratio

$$
\begin{aligned}
\mathrm{S} &= \frac{\text{temporal variation}}{\text{convection}} \\[2ex]
&= \frac{\text{residence time}}{\text{characteristic time}}
\end{aligned}
$$

If

- $\mathrm{S} \gg 1$ then, the temporal term of the equation is important and cannot be neglected. It is a transient process.
- $\mathrm{S} \ll 1$ then, the flow is steady and the temporal term can be eliminated.

Reynolds number

The Reynolds number stems from dividing the convective term by the viscous diffusion term, representing the ratio of inertial to viscous forces,

$$\text{Re} = \frac{\text{convection}}{\text{diffusion}} \approx \frac{\text{inertial forces}}{\text{viscous forces}}$$

Thus, if

- Re $<<$ 1, the viscous forces (friction) are dominant. The convective term can be neglected compared to the viscous term.
- Re $>>$ 1, convection (or inertial) forces are dominant. But one should be really cautious here: in this case, the viscous forces cannot be neglected everywhere in the flow, only away from interfaces. In particular, in areas close to interfaces and boundary layers, the viscous forces are of the same order as the inertial forces, because in the vicinity of interfaces, very strong gradients may exist.

Euler Number

The Euler number comes from the ratio between pressure forces and inertial forces,

$$\text{Eu} = \frac{\text{pressure forces}}{\text{inertial forces}}$$

In general this term should be kept in the equations because the pressure forces typically are of the same order as the largest term in the equation (with a few exceptions).

Froude Number

Similarly, the Froude number is the ratio

$$\text{Fr} = \frac{\text{inertial forces}}{\text{body forces}}$$

If

- Fr $>>$ 1, the body forces can be neglected.
- Fr $<<$ 1, the body forces are significant. This number is important for flows with a free surface, such as in the dynamics between the waves and a boat.

Péclet and Péclet II Numbers

The Péclet and Péclet II numbers are analogous to the Reynolds number but for heat and mass transfer, respectively.

The Péclet number is the ratio,

$$Pe = \frac{\text{heat transfer by convection}}{\text{heat transfer by diffusion}}$$

Using the Prandtl number, defined as

$$Pr = \frac{\mu_0 c_{p0}}{\kappa_0} = \frac{\text{viscous diffusion}}{\text{thermal diffusion}}$$

we can write the Péclet number as a function of Reynolds number,

$$\boxed{Pe = Re\ Pr}$$

The Prandtl number indicates the importance of momentum transport by diffusion compared to heat transport by diffusion.

Likewise, for mass transfer,

$$Pe_{II} = \frac{\text{mass transfer of component } A \text{ by convection}}{\text{mass transfer of component } A \text{ by diffusion}}$$

This number compares mass transport phenomena by convection and diffusion.

Other relevant dimensionless numbers are

$$Sc = \frac{\mu_0}{\rho_0 D_0} = \frac{\text{viscous diffusion}}{\text{mass diffusion}}$$

$$Le = \frac{\kappa_0}{\rho_0 C_{p0} D_0} = \frac{\text{thermal diffusion}}{\text{mass diffusion}}$$

As a function of these numbers, we can write

$$\boxed{\begin{aligned} Pe_{II} &= Re\ Sc \\[2mm] Pe_{II} &= Pe\ Le \end{aligned}}$$

The numbers Prandtl, Schmidt and Lewis are properties of the fluid. For instance, for air $Pr = 0.7$ and for water, about $Pr = 5$, a value that depends on temperature (see Table 10.1).

Damköhler Number

The importance of the chemical production of component A compared to the convective term is given by the Damköhler number,

$$Da_I = \frac{\text{production of component } A}{\text{convection}}$$

If

- $Da_I \ll 1$, the chemical production can be ignored. The chemical reaction is slow compared to the convective time.
- $Da_I \gg 1$, the chemical production is important. The reaction is fast, and equilibrium conditions can be assumed.
- $Da_I \approx 1$, the chemical time is of the same order as the convective time. For a precise calculation, the chemical kinetics should be modeled.

Table 10.1. Ranges of Prandtl number, Pr, for various substances [27].

Fluid	Pr
Liquid metals	$0.004 - 0.03$
Gases	$0.7 - 1.0$
Water	$1.7 - 13.7$
Light organic liquids	$5 - 50$
Oils	$50 - 10\,000$
Glycerin	$2\,000 - 100\,000$

Problems

10.1 The transport equation that governs the concentration of suspended sediments S [kg/m^3] in a stream of depth h is

$$\frac{\partial hS}{\partial t} + \frac{\partial uhS}{\partial x} + \frac{\partial vhS}{\partial y} = \frac{\partial}{\partial x}\left(\epsilon h \frac{\partial S}{\partial x}\right) + \frac{\partial}{\partial y}\left(\epsilon h \frac{\partial S}{\partial y}\right) + R - w_f S$$

where h is the water level; u, v velocity components; R the rate of sediment re-suspension; and w_f the sedimentation factor.

(a) Determine the dimensions of ϵ, R, w_f.
(b) Find the dimensionless numbers that characterize the sediment transport.
(c) Identify the dimensionless numbers.

10.2 In the previous exercise, give criteria for:

(a) Steady flow.
(b) Negligible viscous forces.
(c) Important sedimentation rate.
(d) Apply the three above conditions for the case $h = 1$ m; $U = 0.5$ m/s; $\epsilon = 0.001$; $w_f = 0.1$; characteristic time $= 1$ day.

10.3 The salt concentration of sea water obeys the following relation:

$$\frac{\partial S}{\partial t} + u\frac{\partial S}{\partial x} + v\frac{\partial S}{\partial y} + w\frac{\partial S}{\partial z} = K\left(\frac{\partial^2 S}{\partial x^2} + \frac{\partial^2 S}{\partial y^2} + \frac{\partial^2 S}{\partial z^2}\right)$$

where K is the mass diffusion coefficient, of units m^2/s and S, the salinity (parts per thousand). Make the equation nondimensional and discuss the encountered parameters.

10.4 In a porous media, the combination of the continuity equation,

$$\epsilon\frac{\partial \rho}{\partial t} = \boldsymbol{\nabla} \cdot (\rho \boldsymbol{v})$$

and the Darcy law,

$$v = \frac{\gamma}{\mu}\left(\nabla p - \rho \boldsymbol{g}\right)$$

gives rise to the equation

$$\frac{\epsilon\mu}{\gamma}\frac{\partial\rho}{\partial t} = \nabla \cdot \left[\rho\left(\nabla p - \rho\boldsymbol{g}\right)\right]$$

where γ is the medium permeability $[\text{m}^2]$, ϵ the medium porosity, and the remaining symbols have the usual meaning.

(a) Determine the dimensions of ϵ.

(b) Nondimensionalize the equation and obtain the relevant dimensionless parameters.

10.5   The partial differential equation that governs the energy transport in a two-dimensional incompressible porous media can be approximated as:

$$-\rho c_p \frac{\gamma}{\mu}\frac{\partial p}{\partial x}\frac{\partial T}{\partial x} - \rho c_p \frac{\gamma}{\mu}\frac{\partial p}{\partial y}\frac{\partial T}{\partial y} = \kappa\frac{\partial^2 T}{\partial x^2} + \kappa\frac{\partial^2 T}{\partial y^2}$$

where γ is a property of the porous medium called *permeability*. The rest of the parameters have the usual meaning.

(a) What are the dimensions of γ?

(b) In order to nondimensionalize the equation, besides using the properties of the medium γ, μ and c_p, employ the characteristic scales (l_0, v_0, p_0, T_0). Discuss the resulting dimensionless parameters.

10.6   In natural convection flow problems, the variation of density due to small temperature differences ΔT can be modeled with the Boussinesq approximation. This approximation incorporates the buoyancy forces into the momentum equations, resulting in

$$\rho_0\frac{D\boldsymbol{v}}{Dt} = -\nabla(p + \rho_0 g z) - \rho_0\beta\Delta T\boldsymbol{g} + \mu\nabla^2\boldsymbol{v}$$

where ρ_0 is the fluid density at the reference temperature, T_0; $\Delta T = T - T_0$ is the temperature difference causing the motion and β, the thermal expansion coefficient. Find the relevant dimensionless parameters.

10.7   The momentum equation for a rotating frame of reference, with an angular velocity $\boldsymbol{\omega}$, can be written as follows

$$\rho\frac{\partial\boldsymbol{v}}{\partial t} + \rho\boldsymbol{v}\cdot\nabla\boldsymbol{v} = -\nabla p + \mu\Delta\boldsymbol{v} - \underbrace{\rho\,\boldsymbol{\omega}\times\boldsymbol{\omega}\times\boldsymbol{r}}_{\text{centrifugal force}} - \underbrace{\rho\,2\,\boldsymbol{\omega}\times\boldsymbol{v}}_{\text{Coriolis}} + \rho\boldsymbol{f}_m$$

where $\boldsymbol{r}$ is the position vector with respect to the rotation axis; $\boldsymbol{v}$, the velocity vector; Δ represents here the Laplacian, $\Delta = \nabla\cdot\nabla = \nabla^2$; and the rest of the variables have the usual meaning.

(a) Find the dimensionless numbers that indicate the relevance of the centrifugal and Coriolis forces.
(b) Identify the rest of the dimensionless numbers.

10.8 The fluid flow in porous media, such as the flow in soils of sand and clay, can be described by the Brinkman equation,

$$-\nabla p - \frac{\mu}{\gamma} \boldsymbol{v} + \mu \nabla^2 \boldsymbol{v} + \rho \boldsymbol{g} = 0$$

where p is the pressure, $\boldsymbol{v}$ the velocity vector, μ the dynamic viscosity, γ the permeability of the medium, ρ the density and $\boldsymbol{g}$ the vector of gravitational acceleration.

(a) Determine the dimensions of γ.
(b) Nondimensionalize the equation and obtain the relevant dimensionless parameters.

Part IV

Transport Phenomena at Interfaces

Introduction to the Boundary Layer

Boundary layers are of vital importance for transport phenomena. They are thin areas within the fluid, with large property gradients, that appear in solid/fluid and fluid/fluid interfaces. Despite being a small area of the fluid flow, most of the transport processes of momentum, heat and mass take place there. Thus, they are decisive in the design of equipment that involves transport phenomena.

11.1 Concept of Boundary Layer

In engineering we are interested in calculating friction, heat and mass transfer within a fluid. Most of these processes take place at interfaces.

There exist two types of fluid interfaces.

(a) Those in which the fluid is in contact with a solid wall, called solid/fluid interfaces.
(b) Those where fluids with different properties meet, which are called fluid/fluid interfaces. These can be of various types, i.e. gas/gas, gas/liquid or liquid/liquid.

Definition 11.1 (Boundary layer). *The boundary layer is the area around an interface (solid/fluid, fluid/fluid) where there exist large property gradients in the direction across the interface, giving rise to the processes of friction, heat transfer and mass transfer.*

For each transport phenomenon there is a specific boundary layer, with different properties, such as its thickness. According to the transported property involved, boundary layers can be classified as follows.

(a) Viscous boundary layer (for the velocity or momentum), with thickness δ.
(b) Thermal boundary layer (for the temperature or energy), with thickness δ_T.
(c) Mass boundary layer (for the chemical concentrations), with thickness δ_c.

11.2 Laminar versus Turbulent Boundary Layer

As for the flow field, boundary layers can be laminar or turbulent. It is very important to distinguish these two types of flows, because transport phenomena at interfaces depend heavily on this trait. For each kind of flow regime, different correlations are applicable to quantify the transport coefficients.

Laminar flow is typically orderly, predictable and deterministic. Turbulent flow is always three-dimensional, unsteady and chaotic. In turbulence flow, there are random fluctuations of the fluid field, which enhance tremendously the convective transport of properties.

The number that governs the transition from laminar to turbulent flow is the *transition* Reynolds number Re_{tr}, which depends on the geometry of the flow among other factors such as the free-stream turbulence and surface roughness.

11.3 The Prandtl Theory

Taking into account the qualitative structure of the boundary layer, the transport equations can be simplified within interfaces. This is known as *Prandtl's boundary layer theory* (1904), which states that for high Reynolds numbers, the boundary layer thickness δ at a position $x = L$ from the beginning of the boundary layer is much smaller than L, i.e.

$$\boxed{\delta << L} \tag{11.1}$$

Furthermore, inside the boundary layer the convective and diffusion terms are of the same size.

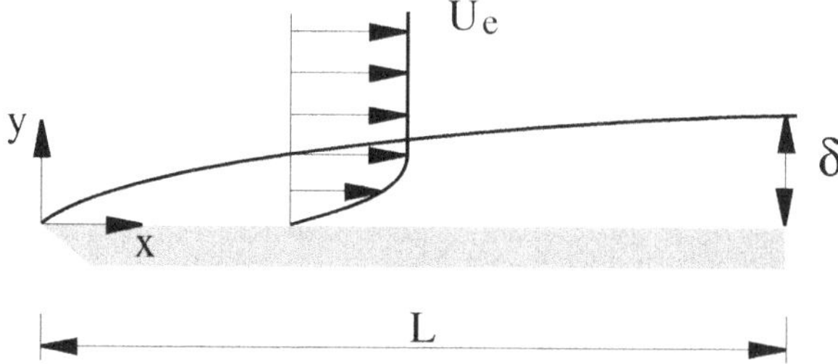

Fig. 11.1. Schematic of the viscous boundary layer.

11.3.1 Estimation of the Boundary Layer Thicknesses for Laminar Flow

Let us choose the coordinate axis shown in Fig. 11.1. The x axis is aligned to the boundary layer along the interface; the y axis is set perpendicularly to the interface.

Viscous Boundary Layer

Due to its structure, inside the boundary layer the x-derivatives are much smaller than the y derivatives. For example, for the viscous boundary layer,

$$\left|\frac{\partial u}{\partial x}\right| << \left|\frac{\partial u}{\partial y}\right| \tag{11.2}$$

$$\left|\frac{\partial^2 u}{\partial x^2}\right| << \left|\frac{\partial^2 u}{\partial y^2}\right| \tag{11.3}$$

This can be verified by approximating the derivatives as Fig. 11.2 shows, which gives the following orders of magnitude:

$$\frac{\partial u}{\partial x} \approx \frac{U_e - 0}{L} \tag{11.4}$$

$$\frac{\partial^2 u}{\partial x^2} \approx \frac{U_e - 0}{L^2} \tag{11.5}$$

$$\frac{\partial u}{\partial y} \approx \frac{U_e - 0}{\delta} \tag{11.6}$$

$$\frac{\partial^2 u}{\partial y^2} \approx \frac{U_e - 0}{\delta^2} \tag{11.7}$$

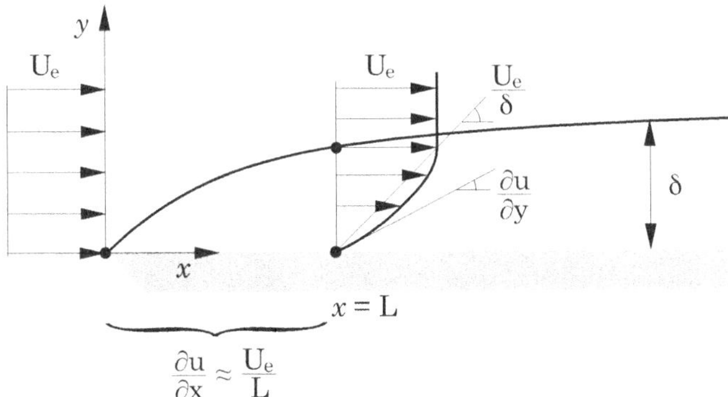

Fig. 11.2. Estimating velocity derivatives inside the viscous boundary layer.

Since within the boundary layer the convective term is of the same order of magnitude as the viscous term,

$$u\frac{\partial u}{\partial x} \approx \nu\frac{\partial^2 u}{\partial y^2} \qquad (11.8)$$

Approximating the derivatives as above,

$$U_e\frac{U_e - 0}{L} \approx \nu\frac{U_e - 0}{\delta^2} \qquad (11.9)$$

and simplifying

$$\left(\frac{\delta}{L}\right)^2 \approx \frac{\nu}{U_e L}$$

one arrives at

$$\boxed{\frac{\delta}{L} \approx \frac{1}{\sqrt{\mathrm{Re}_L}}} \qquad (11.10)$$

Note that the Reynolds number is based on the length L, and consequently, it is denoted as Re_L. The Reynolds number is typically based on a characteristic dimension of the object under analysis. For instance, if we had a uniform flow around a cylinder, the characteristic length could be taken as the cylinder diameter D and the Reynolds number, Re_D.

Remark 11.1. For a given position $x = L$, if the Reynolds number increases, the boundary layer thickness decreases. For instance, for a given free-stream velocity U_e and length L, the fluid with the smallest kinematic viscosity ν will have the smallest boundary layer thickness δ.

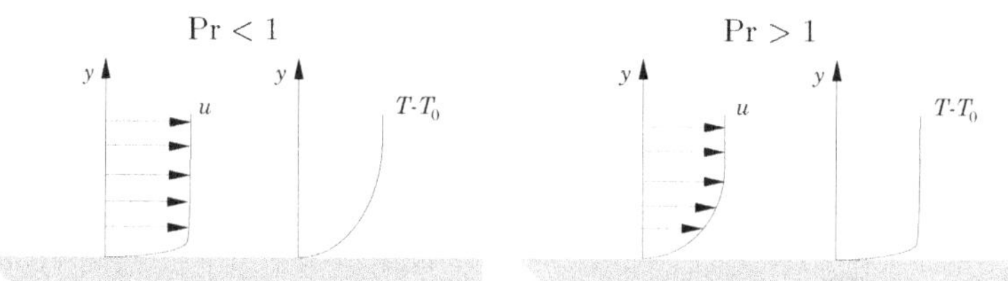

Fig. 11.3. Velocity and temperature distributions at the same coordinate of the viscous and thermal boundary layers for Pr < 1 and Pr > 1.

Thermal Boundary Layer

For the thermal boundary layer, two cases can be distinguished: $Pr \ll 1$ and $Pr \gg 1$ (see Fig. 11.3).

- $Pr \ll 1$. In this case, $\delta \ll \delta_T$ and the velocity scale within the thermal boundary layer can be taken as U_e. Imposing that in the thermal boundary layer, the convective term is of the same order as the viscous term,

$$u\frac{\partial T}{\partial x} \approx \alpha\frac{\partial^2 T}{\partial y^2} \tag{11.11}$$

And approximating the derivatives as above with ΔT the temperature difference across the thermal boundary layer,

$$U_e\frac{\Delta T}{L} \approx \alpha\frac{\Delta T}{\delta_T^2} \tag{11.12}$$

results in

$$\boxed{\frac{\delta_T}{L} \approx \frac{1}{\sqrt{Pe_L}}} \tag{11.13}$$

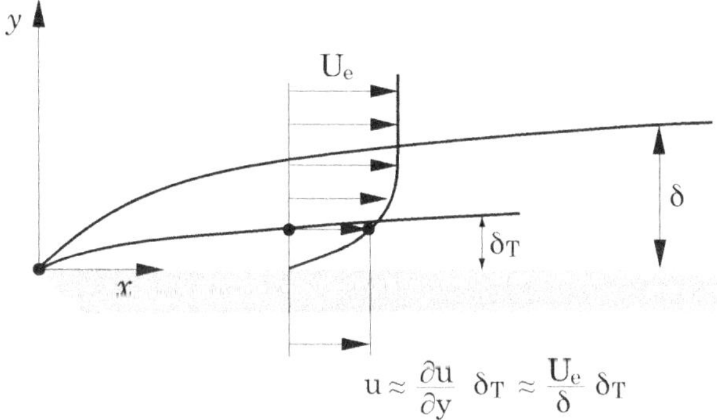

Fig. 11.4. Estimating the characteristic velocity for the thermal boundary layer at $Pr \gg 1$.

- $Pr \gg 1$, which implies $\delta \gg \delta_T$. In this case, since the viscous boundary layer is much thicker than the thermal layer, the characteristic velocity within the thermal boundary layer can be taken as the velocity at $y = \delta_T$ (see Fig. 11.4). Assuming a linear near-wall velocity profile in the viscous boundary layer, the characteristic velocity can be taken as

$$u(y = \delta_T) \approx \frac{\delta_T}{\delta}U_e \tag{11.14}$$

Again, considering that thermal convection is as important as thermal diffusion,

$$u\frac{\partial T}{\partial x} \approx \alpha\frac{\partial^2 T}{\partial y^2} \tag{11.15}$$

and substituting the above orders of magnitude

$$\frac{\delta_T}{\delta}U_e\frac{\Delta T}{L} \approx \alpha\frac{\Delta T}{\delta_T^2} \tag{11.16}$$

which simplifies to

$$\boxed{\frac{\delta_T}{L} \approx \mathrm{Re}^{-1/2}\mathrm{Pr}^{-1/3} = \mathrm{Pe}^{-1/2}\mathrm{Pr}^{1/6}} \tag{11.17}$$

Mass Boundary Layer

For the mass boundary layer, similar conclusions as for the thermal boundary layer are reached, but substituting Pe_L by Pe_{II_L} and Pr by Sc:

- Sc $\ll$ 1

$$\boxed{\frac{\delta_c}{L} \approx \frac{1}{\sqrt{\mathrm{Pe}_{II_L}}}} \tag{11.18}$$

- Sc $\gg$ 1

$$\boxed{\frac{\delta_c}{L} \approx \mathrm{Re}_L^{-1/2}\mathrm{Sc}^{-1/3} = \mathrm{Pe}_{II_L}^{-1/2}\mathrm{Sc}^{1/6}} \tag{11.19}$$

Remark 11.2. Laminar boundary layers grow with the square root of the distance,

$$\delta \approx L^{1/2} \qquad \delta_T \approx L^{1/2} \qquad \delta_c \approx L^{1/2}$$

11.3.2 Relative Boundary Layer Thicknesses

Combining the above approximations for the boundary layer thicknesses, the following conclusions are reached.

$$\mathrm{Pr} \ll 1 \quad \frac{\delta}{\delta_T} \approx \mathrm{Pr}^{-1/2} \tag{11.20}$$

$$\mathrm{Pr} \gg 1 \quad \frac{\delta}{\delta_T} \approx \mathrm{Pr}^{-1/3} \tag{11.21}$$

$$\mathrm{Sc} \ll 1 \quad \frac{\delta}{\delta_c} \approx \mathrm{Sc}^{-1/2} \tag{11.22}$$

$$\mathrm{Sc} \gg 1 \quad \frac{\delta}{\delta_c} \approx \mathrm{Sc}^{-1/3} \tag{11.23}$$

11.4 Incompressible Boundary Layer Equations

The equations at the boundary layer are derived from the conservation equations taking into account the physical structure of these layers.

For the sake of simplicity, let us assume that the flow is two-dimensional, incompressible, steady, with constant fluid properties (ρ, μ, κ, D_{AB}, c_p, etc.), negligible viscous dissipation (small Mach number) and without chemical reaction ($\dot{\omega}_A = 0$). Under the above hypothesis, the equations of continuity, momentum, temperature and chemical concentration become

$$\frac{\partial u}{\partial x} + \frac{\partial v}{\partial y} = 0 \tag{11.24}$$

$$u\frac{\partial u}{\partial x} + v\frac{\partial u}{\partial y} = -\frac{1}{\rho}\frac{\partial p}{\partial x} + \nu\left(\frac{\partial^2 u}{\partial x^2} + \frac{\partial^2 u}{\partial y^2}\right) \tag{11.25}$$

$$u\frac{\partial v}{\partial x} + v\frac{\partial v}{\partial y} = -\frac{1}{\rho}\frac{\partial p}{\partial y} + \nu\left(\frac{\partial^2 v}{\partial x^2} + \frac{\partial^2 v}{\partial y^2}\right) \tag{11.26}$$

$$u\frac{\partial T}{\partial x} + v\frac{\partial T}{\partial y} = +\alpha\left(\frac{\partial^2 T}{\partial x^2} + \frac{\partial^2 T}{\partial y^2}\right) \tag{11.27}$$

$$u\frac{\partial c_A}{\partial x} + v\frac{\partial c_A}{\partial y} = +D_A\left(\frac{\partial^2 c_A}{\partial x^2} + \frac{\partial^2 c_A}{\partial y^2}\right) \tag{11.28}$$

where

$$\nu = \frac{\mu}{\rho} \qquad \alpha = \frac{\kappa}{\rho c_p}$$

are, respectively, the kinematic viscosity and the thermal diffusivity. Recall that ν, α and D_A have the same dimensions, L^2/T.

These transport equations can be simplified studying the *orders of magnitude* of all the terms, as in (11.9), to finally yield the boundary layer equations.

11.4.1 Continuity Equation

The continuity equation gives us an estimate of the vertical velocity component V_e inside the boundary layer.

The orders of magnitude of the two terms of the continuity equation,

$$\frac{\partial u}{\partial x} + \frac{\partial v}{\partial y} = 0 \tag{11.29}$$

can be estimated as,

$$\frac{U_e - 0}{L} + \frac{V_e - 0}{\delta} \approx 0 \tag{11.30}$$

Simplifying, one gets

$$\boxed{V_e \approx \frac{\delta}{L}U_e} \tag{11.31}$$

This equation informs us that inside the boundary layer, the v component of the velocity is much smaller than the u component.

11.4.2 x-Momentum Equation

For this equation, the convective terms can be estimated as

$$u\frac{\partial u}{\partial x} \approx U_e\frac{U_e}{L} = \frac{U_e^2}{L} \tag{11.32}$$

$$v\frac{\partial u}{\partial y} \approx V_e\frac{U_e}{\delta} = \frac{U_e^2}{L} \tag{11.33}$$

where the order of magnitude of v derived from the continuity equation has been used. This result implies that both convective terms are of the same order and none can be ignored.

The size of the viscous terms can be approximated as

$$\nu\frac{\partial^2 u}{\partial x^2} \approx \nu\frac{U_e}{L^2} \tag{11.34}$$

$$\nu\frac{\partial^2 u}{\partial y^2} \approx \nu\frac{U_e}{\delta^2} \tag{11.35}$$

As a consequence, inside the boundary layer,

$$\nu\frac{\partial^2 u}{\partial x^2} \approx \nu\frac{\partial^2 u}{\partial y^2}\left(\frac{\delta}{L}\right)^2 \tag{11.36}$$

and, therefore,

$$\boxed{\nu\frac{\partial^2 u}{\partial x^2} << \nu\frac{\partial^2 u}{\partial y^2}} \tag{11.37}$$

so the term on the left-hand side can be ignored compared to the right-hand side term.

The pressure forces, in principle, cannot be ignored. Thus, they should be of the same order as the largest term in the x-momentum equation,

$$\frac{1}{\rho}\frac{\Delta_x p}{L} \approx \frac{U_e^2}{L} \tag{11.38}$$

and, therefore, the x-variation of the pressure must be of the order of

$$\boxed{\Delta_x p \approx \rho U_e^2} \tag{11.39}$$

11.4.3 y-Momentum Equation

Likewise, the terms in this equation can be estimated as follows,

$$u\frac{\partial v}{\partial x} \approx U_e\frac{V_e}{L} = \frac{U_e^2}{L}\frac{\delta}{L} \tag{11.40}$$

$$v\frac{\partial v}{\partial y} \approx V_e\frac{V_e}{\delta} = \frac{U_e^2}{L}\frac{\delta}{L} \tag{11.41}$$

$$\nu\frac{\partial^2 v}{\partial x^2} \approx \nu\frac{V_e}{L^2} = \nu\frac{U_e}{L^2}\frac{\delta}{L} \tag{11.42}$$

$$\nu\frac{\partial^2 v}{\partial y^2} \approx \nu\frac{V_e}{\delta^2} = \nu\frac{U_e}{\delta^2}\frac{\delta}{L} \tag{11.43}$$

One can notice that the y-momentum equation is δ/L times smaller than the x-counterpart. Similarly, the pressure gradient has to be of the same order as the largest term in this equation,

$$\frac{1}{\rho}\frac{\Delta_y p}{\delta} \approx \frac{U_e^2}{L}\frac{\delta}{L} \tag{11.44}$$

and, therefore,

$$\Delta_y p \approx \Delta_x p \left(\frac{\delta}{L}\right)^2 \tag{11.45}$$

As a consequence, the vertical pressure gradient can be neglected,

$$\boxed{\frac{\partial p}{\partial y} \approx \frac{\partial p}{\partial x}\frac{\delta}{L} \to 0} \tag{11.46}$$

implying that the pressure is constant across the boundary layer and $p = p(x)$. Finally, the y-momentum equation is replaced by (11.46).

11.4.4 Temperature and Concentration Equations

The analysis above can be repeated for the temperature and concentration equations. We will do it for a general scalar ϕ, with equation

$$u\frac{\partial \phi}{\partial x} + v\frac{\partial \phi}{\partial y} = \alpha_\phi\left(\frac{\partial^2 \phi}{\partial x^2} + \frac{\partial^2 \phi}{\partial y^2}\right) \tag{11.47}$$

We will denote by $\Delta\phi$ the variation of the scalar across the boundary layer,

$$\Delta\phi = \phi_0 - \phi_\infty \tag{11.48}$$

where ϕ_0 is the value of the scalar at the solid boundary and ϕ_∞, the value at the free-stream. For this scalar, the boundary layer thickness is denoted by δ_ϕ.

Then, the size of the the convective terms can be estimated as

$$u\frac{\partial \phi}{\partial x} \approx U_e\frac{\Delta\phi}{L} = U_e\frac{\Delta\phi}{L} \tag{11.49}$$

$$v\frac{\partial \phi}{\partial y} \approx V_e\frac{\Delta\phi}{\delta_\phi} = U_e\frac{\delta}{\delta_\phi}\frac{\Delta\phi}{L} \approx U_e\frac{\Delta\phi}{L} \tag{11.50}$$

Above it has been assumed that the thickness of the momentum, energy and mass are of the same order (but not equal), i.e. $\delta_\phi \approx \delta$. As before, both convective terms are of the same order and should be retained.

Similarly the viscous terms can be estimated as

$$\alpha_\phi \frac{\partial^2 \phi}{\partial x^2} \approx \alpha_\phi \frac{\Delta\phi}{L^2} \tag{11.51}$$

$$\alpha_\phi \frac{\partial^2 \phi}{\partial y^2} \approx \alpha_\phi \frac{\Delta\phi}{\delta_\phi^2} \tag{11.52}$$

and, therefore,

$$\alpha_\phi \frac{\partial^2 \phi}{\partial x^2} \approx \alpha_\phi \frac{\partial^2 \phi}{\partial y^2} \left(\frac{\delta_\phi}{L}\right)^2 \tag{11.53}$$

Thus, as for the viscous boundary layer,

$$\boxed{\alpha_\phi \frac{\partial^2 \phi}{\partial x^2} << \alpha_\phi \frac{\partial^2 \phi}{\partial y^2}} \tag{11.54}$$

and the first term can be neglected compared to the second term.

11.4.5 Boundary Layer Equations: Summary

Gathering all the equations derived in the above section, the incompressible, two-dimensional, steady, constant diffusion coefficient, boundary layer equations can be expressed as follows.

$$\frac{\partial u}{\partial x} + \frac{\partial v}{\partial y} = 0 \tag{11.55}$$

$$u\frac{\partial u}{\partial x} + v\frac{\partial u}{\partial y} = -\frac{1}{\rho}\frac{\partial p}{\partial x} + \nu\left(\frac{\partial^2 u}{\partial y^2}\right) \tag{11.56}$$

$$0 = -\frac{1}{\rho}\frac{\partial p}{\partial y} \tag{11.57}$$

$$u\frac{\partial T}{\partial x} + v\frac{\partial T}{\partial y} = +\alpha\left(\frac{\partial^2 T}{\partial y^2}\right) \tag{11.58}$$

$$u\frac{\partial c_A}{\partial x} + v\frac{\partial c_A}{\partial y} = +D_A\left(\frac{\partial^2 c_A}{\partial y^2}\right) \tag{11.59}$$

For laminar flow and simple geometries, there are semi-analytical solutions for these equations which must be expressed in tabular and graphical form. The first solution of the momentum boundary layer for a flat plate is due to Blasius and was obtained assuming self-similarity for the flow variables inside the boundary layer (see, for instance [21, 26]).

11.5 Measures of the Boundary Layer Thickness

The boundary layer is the area in the neighborhood of the interfaces, where the fluid variables evolve from the interface to the free-stream values. Rigorously speaking, for the viscous boundary layer, its thickness $\delta(x)$ is defined as

$$u(x, y = \delta(x)) = U_e(x) \tag{11.60}$$

and analogously for the thermal and mass boundary layers,

$$\begin{aligned}
T(x, y = \delta_T(x)) &= T_e(x) \\
c_A(x, y = \delta_c(x)) &= c_{Ae}(x)
\end{aligned} \tag{11.61}$$

However, other definitions for the boundary layer thickness, linked to different physical phenomena, are possible.

Thickness 99%

Sometimes, it might be very difficult to measure or compute where condition (11.60) is met. In order to facilitate its measurement or computation, this criterion is relaxed and the boundary layer thickness is defined as the distance from the interface where the variable reaches the 99% free-stream local value.

Note that this percentage is arbitrary, making it possible to find in practice other values, (such as 90%), even though 99% is the most usual. To specify which definition of boundary layer thickness is been used, the percentage is typed as a subscript of δ.

Thus, the δ_{99} boundary layer thickness is defined for the viscous boundary layer as

$$u(x, y = \delta_{99}(x)) = 0.99\, U_e \tag{11.62}$$

and for the thermal and mass boundary layers,

$$\begin{aligned}
T(x, y = \delta_{T99}(x)) - T_0 &= 0.99\,(T_e - T_0) \\
c_A(x, y = \delta_{c99}(x)) - c_{A0} &= 0.99\,(c_{Ae} - c_{A0})
\end{aligned} \tag{11.63}$$

Integral Measures

There are other measures of the boundary layer thickness, like the displacement thickness, momentum thickness, enthalpy thickness and concentration thickness, that appear in the integral boundary layer equations. These are, however, out of the scope of this text.

Problems

11.1 Verify the order of magnitude of δ, δ_T and δ_c for laminar flow.

11.2 Check the expressions derived for the relative boundary layer thicknesses of Section 11.3.2.

Momentum, Heat and Mass Transport

Most of the transport processes occur near thin interfaces and boundary layers. Since fluid flow is governed by a coupled nonlinear system of partial differential equations, the calculation of transport processes in industrial applications can be quite a costly task. However, in many practical situations, the calculation of friction or energy losses in conduits, or the heat and mass transport at interfaces can be greatly simplified by using the so-called transport coefficients. These coefficients, which are based on analytical solutions or experimental data, are typically gathered as correlations based on dimensionless numbers, ready for use.

12.1 The Concept of Transport Coefficient

Assume it is desired to determine the transport of momentum, heat or mass at an interface (solid/fluid or fluid/fluid). According to the constitutive equations, for an interface parallel to the x axis, placed at $y = 0$, the diffusion fluxes across the interface are given by

$$\tau' = \mu \left.\frac{\partial u}{\partial y}\right|_{y=0} \qquad q = -\kappa \left.\frac{\partial T}{\partial y}\right|_{y=0} \qquad j_A = -D_A \left.\frac{\partial \rho_A}{\partial y}\right|_{y=0}$$

Therefore, to calculate the flux of any transport phenomena at an interface, it is necessary to determine not only the fluid field, but the derivatives of the velocity, temperature and chemical concentrations, too. These not only can be quite costly to obtain, but sometimes, even impossible. On the one side, in fluid mechanics there are only a few analytical solutions, and always for simple geometries. On the other side, when the flow is turbulent, reliable numerical computation of industrial configurations is typically out of reach for today's computers.

The Local Transport Coefficient

Frequently, instead of very accurate values of the transport fluxes, a good approximation based on empirical correlations can be sufficient. In particular, instead of using the constitutive equations, the flux of a property across an interface can be approximated from the algebraic relation

$$\boxed{\,\mathrm{d}\Gamma_0 = K\,(\Pi_0 - \Pi_\infty)\,\mathrm{d}A\,}$$

(12.1)

In this expression, Γ_0 is the property flux at the interface, being positive when towards the flow; Π_0 is the property value at the interface; Π_∞ is the reference value of the property, generally the free-stream value; $\mathrm{d}A$, the elemental differential of area; and K, the *local transport coefficient*.

Therefore, the transport coefficient K relates a characteristic difference of the variable $(\Pi_0 - \Pi_\infty)$ with the flux Γ_0, allowing its computation without knowledge of the derivative in a simple and practical way. Examples of transport coefficients are found in the table below.

Table 12.1. Transport coefficient for various phenomena.

Flux	Γ	$\mathrm{d}\Gamma/\mathrm{d}A$	Π	Transport Coefficient
Force	F [N]	τ_0' [Pa]	ρV	K_τ
Heat/time	$\dot{Q}$ [W]	q [W/m^2]	T	h
Mass of A/time	J_A [kg/s]	j_A [kg/(m^2 s)]	ρ_A	h_m

The transport coefficient is generally determined from experiments. And, whereas the constitutive relations are general, the transport coefficient depends on the flow configuration, i.e. the geometrical shape, the boundary conditions, whether the flow is laminar or turbulent, etc.

Note that, in the definition of the transport coefficient, the derivative of the variable is estimated from the difference of a fluid variable between the interface and free-stream values.

Remark 12.1. Sometimes, the reference value Π_∞ is taken at some other point of the flow or takes on a more complex physical definition.

The Global Transport Coefficient

The above introduced transport coefficient is the local transport coefficient because it applies to an elemental differential of surface, that is, a local zone.

Likewise, it is possible to define a *global transport coefficient* $\overline{K}$, that gives the flux for the whole area A of interest,

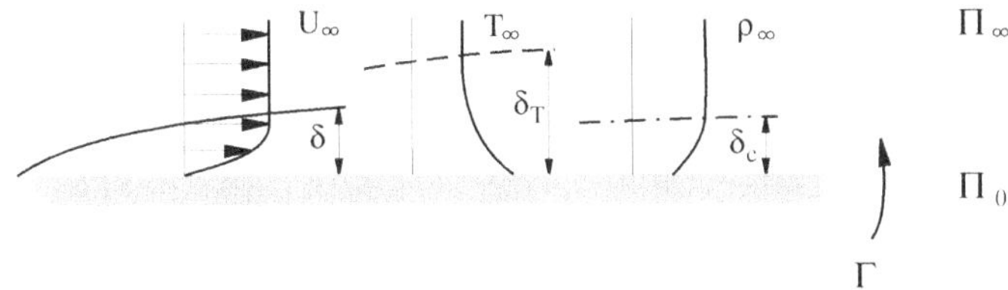

Fig. 12.1. Viscous, thermal and mass boundary layer.

$$\boxed{\Gamma_0 = \overline{K}\,(\Pi_0 - \Pi_\infty)\,A} \tag{12.2}$$

The relation between the global and local transport coefficients can be obtained by integrating the elemental flux $d\Gamma_0$ along the surface A,

$$
\begin{aligned}
\Gamma_0 = \int_A d\Gamma_0 \;&=\; \int_A K\,(\Pi_0 - \Pi_\infty)\,dA \\
&= (\Pi_0 - \Pi_\infty)\int_A K\,dA \equiv \overline{K}\,(\Pi_0 - \Pi_\infty)\,A
\end{aligned}
\tag{12.3}
$$

Thus, from the last line,

$$\boxed{\overline{K} = \frac{1}{A}\int_A K\,dA} \tag{12.4}$$

Many times, the overbar is omitted from the global transport coefficient. In this case, whether the transport coefficient corresponds to the local or global parameter should be distinguished from the context.

12.2 Momentum Transport

The momentum transport coefficient is related to other commonly used parameters to determine drag or pipe losses, like the drag C_D or friction C_f coefficients.

For momentum transport, in the general definition of the transport coefficient we must identify

$$
\begin{aligned}
\Gamma_0 &\longrightarrow F \\
\Pi &\longrightarrow \rho V \\
K &\longrightarrow K_\tau
\end{aligned}
\tag{12.5}
$$

For a still interface, $V_0 = 0$, and

$$\boxed{
\begin{aligned}
dF &= K_\tau\,(\rho V_0 - \rho V_\infty)\,dA \\
&= -K_\tau\,\rho V_\infty\,dA
\end{aligned}
} \tag{12.6}$$

Fig. 12.2. Viscous boundary layer along a flat plate.

Relation Between the Transport and Drag Coefficients

Reorganizing and using the shear stress at the solid boundary, τ_0',

$$K_\tau = -\frac{\dfrac{\mathrm{d}F}{\mathrm{d}A}}{\rho\, V_\infty} = -\frac{-\tau_0'}{\rho\, V_\infty} = \frac{\tau_0'}{\rho\, V_\infty} \tag{12.7}$$

Therefore, the momentum transport coefficient can be calculated from τ_0' at the boundary for the given flow configuration.

The order of magnitude of K_τ is given by

$$K_\tau = \frac{\mu\left.\dfrac{\partial u}{\partial y}\right|_{y=0}}{\rho\, V_\infty} \approx \mathcal{O}\left(\frac{\mu\dfrac{V_\infty}{\delta}}{\rho\, V_\infty}\right) = \mathcal{O}\left(\frac{\mu}{\rho\,\delta}\right) \tag{12.8}$$

where $\mathcal{O}(\cdot)$ denotes *order of magnitude*. This expression shows the influence of the boundary layer thickness δ on the transport of momentum. For the same free-stream velocity and fluid properties, if the boundary layer thickness decreases, the friction increases.

In order to measure the momentum transport coefficient, or to correlate the experimental values, we must apply the theory of dimensional analysis.

The momentum transport coefficient for a steady, incompressible, isothermal flow depends on

$$K_\tau = \varphi(\rho, \mu, V_\infty, L) \tag{12.9}$$

However, since this transport coefficient is related to the wall shear stress, τ_0', it can be reformulated as

$$\tau_0' = \frac{\mathrm{d}F}{\mathrm{d}A} = \varphi(\rho, \mu, V_\infty, L) \tag{12.10}$$

Application of dimensional analysis to the above function (see Chapter 9) gives

$$C_\mathrm{f} = C_\mathrm{f}(\mathrm{Re}) \tag{12.11}$$

where,

$$C_f = \frac{\tau_0'}{\frac{1}{2}\rho V_\infty^2} \tag{12.12}$$

From this expression and that for K_τ (12.7), we have

$$\begin{aligned} \tau_0' &= \tfrac{1}{2}\rho V_\infty^2 C_f \\ \tau_0' &= \rho V_\infty K_\tau \end{aligned} \tag{12.13}$$

Dividing both expressions, one can conclude that

$$K_\tau = \frac{1}{2} V_\infty C_f \tag{12.14}$$

Remark 12.2. The *drag* and *friction coefficients* are momentum transport coefficients.

Remark 12.3. The same conclusion applies to the global transport coefficient, where for this case

$$C_f = \frac{F}{\frac{1}{2}\rho V_\infty^2 A} \tag{12.15}$$

Example 12.1. A researcher is interested in the aerodynamic resistance F_D of spheres. He conducts several experiments in an aerodynamic tunnel at atmospheric conditions ($\rho = 1.2$ kg/m^3, $\mu = 1.8 \times 10^{-5}$ kg/(m s)) with a sphere of a diameter $D = 10$ cm, concluding that

$$F_D = 0.0015 \, V^{1.92}$$

where V is the uniform upstream air velocity.

(a) What is the correct way of presenting the correlation?
(b) Determine the resistance of a $D = 2$ cm sphere at an air speed of $V = 1$ m/s.

Solution. (a) As we have shown above, the dimensionless numbers that govern the resistance of bodies in isothermal low speed conditions are C_D and Re_D. Thus, the correlation could be written as

$$C_D = C \, Re_D^m$$

with C and m constants. Let us find C and m. Expanding the dimensionless equation,

$$\frac{F}{\frac{1}{2}\rho V^2 D^2} = C \left(\frac{\rho V D}{\mu} \right)^m$$

and writing it out in the form of the given correlation,

$$F = C\,\frac{1}{2}\rho V^2 D^2 \left(\frac{\rho V D}{\mu}\right)^m = \left[C\,\frac{1}{2}\rho D^2 \left(\frac{\rho D}{\mu}\right)^m\right] V^{2+m}$$

Matching the power of V in both equations, $2 + m = 1.92$ so $m = -0.08$. Matching the constant,

$$C\,\frac{1}{2}\rho D^2 \left(\frac{\rho D}{\mu}\right)^m = 0.0015$$

from where

$$C = 0.306$$

Thus, the correlation in dimensionless form is

$$C_D = 0.306\,\mathrm{Re}_D^{-0.08}$$

(b) For the small sphere, one can calculate that $\mathrm{Re}_D = 1\,333$ so $C_D = 0.2846$ and

$$F_D = \frac{1}{2}\rho V^2 D^2\,C_D = 6.83 \times 10^{-5}\ \mathrm{N}$$

Table 12.2. Correlations for the friction factor in circular cross-section pipes [25].

Flow regime		Friction factor λ
Laminar	$\mathrm{Re}_D < 2\,300$	$64/\mathrm{Re}_D$
Turbulent	$\mathrm{Re}_D > 2\,300$	$\dfrac{1}{\lambda^{1/2}} = -2.0\log\left(\dfrac{\epsilon/D}{3.7} + \dfrac{2.51}{\mathrm{Re}_D\lambda^{1/2}}\right)$ (Colebrook)
	Fully rough $\dfrac{\mathrm{Re}_D\lambda^{1/2}}{\epsilon/D} > 70$	$\dfrac{1}{\lambda^{1/2}} = -2.0\log\left(\dfrac{\epsilon/D}{3.7}\right)$ (Nikuradse)
	Smooth pipes	$\dfrac{1}{\lambda^{1/2}} = 2.0\log\left(\mathrm{Re}_D\lambda^{1/2}\right) - 0.8$ (Prandtl)

Head Losses in Fluid-Flow Conduits

Linear Losses in Pipes

In the chapter on Dimensional Analysis, we found that the head losses for a steady incompressible flow in a constant section pipe could be expressed by the *Darcy-Weisbach* equation

$$h_f = \frac{\Delta p^*}{\rho g} = \lambda(\frac{\epsilon}{D}, \mathrm{Re}_D)\frac{L}{D}\frac{U^2}{2g}$$

where D is the pipe diameter, L the pipe length, U the mean velocity and ϵ the characteristic length of the roughness. Table 12.2 gathers some correlations for λ. Also, the Moody Diagram depicts graphically the friction factor as a function of the Reynolds number based on the pipe diameter and the relative roughness (see Fig. 12.4).

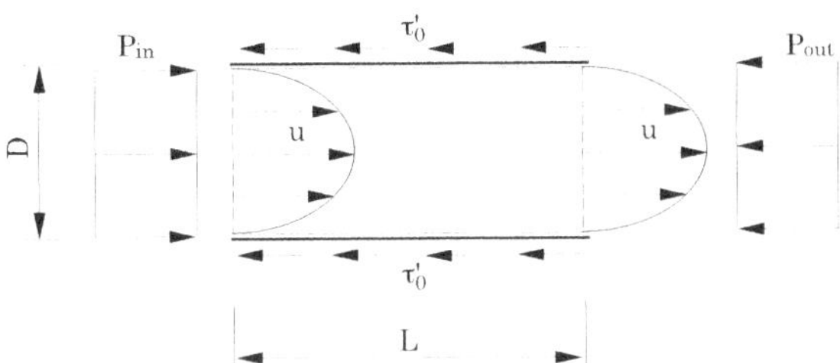

Fig. 12.3. Momentum balance in a horizontal pipe.

A force balance along the pipe axis (see Fig. 12.3) yields that the pressure forces must be equilibrated by the friction forces,

$$\Delta p^* \frac{\pi D^2}{4} = \tau_0' \, \pi D L$$

where $p^* = p + \rho g z$ is *the modified pressure*, which includes the gravity forces. Substituting the value of the pressure loss given by the Darcy-Weisbach equation and τ_0' as a function of C_f,

$$\lambda \frac{L}{D} \frac{1}{2}\rho U^2 \frac{\pi D^2}{4} = C_f \frac{1}{2}\rho U^2 \, \pi D L$$

leads to

$$\boxed{C_f = \frac{\lambda}{4}} \tag{12.16}$$

As a consequence, $\lambda(\frac{\epsilon}{D}, \mathrm{Re}_D)$ is a momentum transport coefficient.

Singular Losses in Pipe Components

In industrial plants, besides constant section straight pipes, one can find components such as bends, valves, tes, expansions and so on. In these elements, the viscous dissipation gives rise to energy losses, which are called *singular* or *local head losses*. These can be computed as

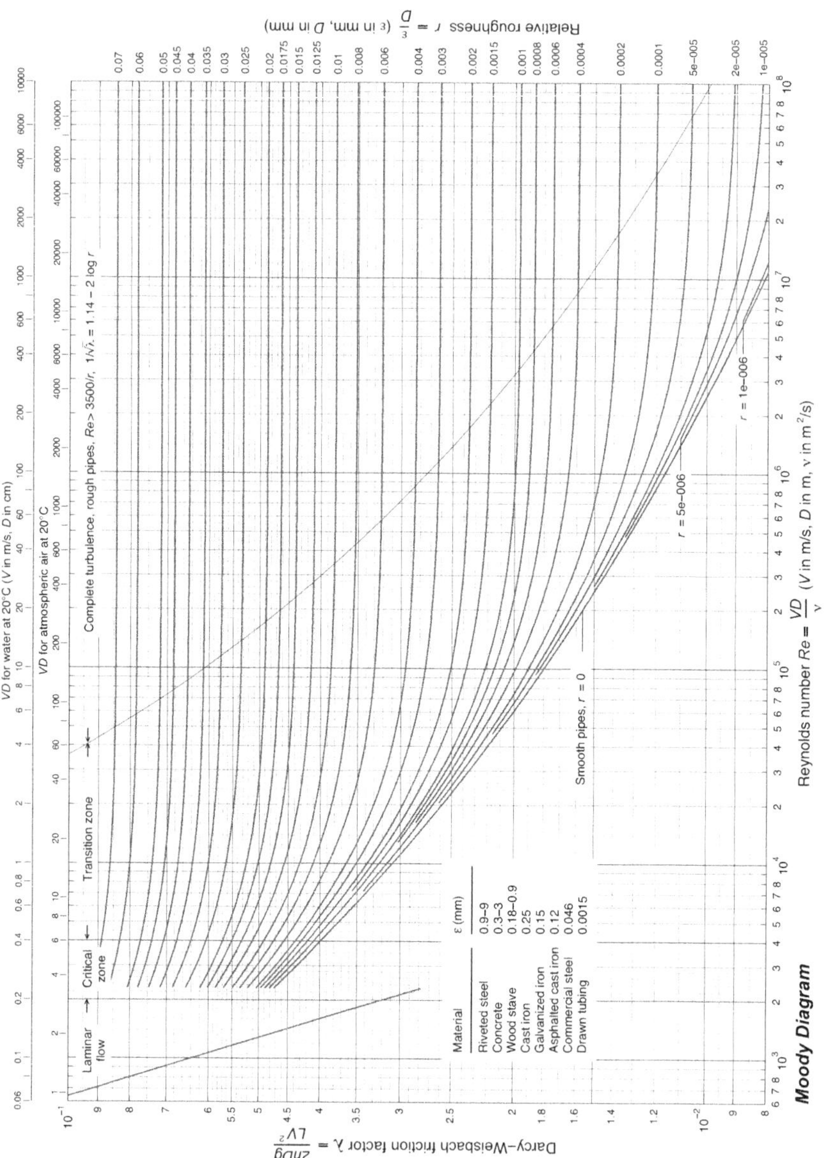

Fig. 12.4. Moody diagram. The Reynolds number is based on the pipe diameter D. Reproduced with permission from [9].

Table 12.3. Guiding values for the local head loss coefficient for several pipe components. The flow goes from section 1 to 2.

Element	K_s
Outlet off tank into pipe	0.05 - 1.
Inlet to tank from pipe	1.
Sudden expansion	$\left(1 - \frac{A_1}{A_2}\right)^2$
Sudden contraction [25]	$\begin{cases} 0.42\left(1 - \frac{A_2}{A_1}\right) & D_2/D_1 \leq 0.76 \\ \left(1 - \frac{A_1}{A_2}\right)^2 & D_2/D_1 > 0.76 \end{cases}$
Long radius $90°$ elbow	0.4
Short radius $90°$ elbow	0.9
Valve (fully opened)	0.03 - 14
Gate valve (fully opened)	0.1
Globe valve (fully opened)	8.0

$$h_f = \frac{\Delta p^*}{\rho g} = K_s \frac{U^2}{2g}$$

where K_s is the dimensionless singular head loss coefficient. The constant K_s depends on the element type and its geometry. Therefore, for each new design, it has to be experimentally determined. Typically, for high Reynolds numbers, K_s is considered independent of Re and so, can be taken as a constant. The reference velocity U can be the upstream or downstream mean velocity. For some examples, see Table 12.3.

As before, if the flow section is the same at the inlet and the outlet of the element, it can be deduced that

$$C_f = \frac{D}{L}\frac{K_s}{4}$$

12.2.1 Basic Momentum Transport Coefficients

The correlations of Table 12.4 are extracted from [16, 10, 3] and are for constant property boundary layers.

12.3 Heat Transport

For the thermal transport coefficient h, we must identify

Table 12.4. Basic momentum transport coefficients.

Geometry	Regime	Correlation
flat plate	laminar $\mathrm{Re}_x < 3 \times 10^5$	$C_{fx} = 0.664 \, \mathrm{Re}_x^{-1/2}$
flat plate	laminar $\mathrm{Re}_x < 3 \times 10^5$	$C_{fL} = 1.328 \, \mathrm{Re}_L^{-1/2}$
flat plate	turbulent $\mathrm{Re}_x > 3 \times 10^5$	$C_{fx} = 0.0576 \, \mathrm{Re}_x^{-1/5}$
pipe	laminar $\mathrm{Re}_D < 2\,300$	$C_f = 16 \, \mathrm{Re}_D^{-1}$
pipe	turbulent $3 \times 10^4 < \mathrm{Re}_D < 10^6$	$C_f = 0.046 \, \mathrm{Re}_D^{-1/5}$

$$\begin{aligned} \Gamma_0 &\longrightarrow \dot{Q} \\ \Pi &\longrightarrow T \\ K &\longrightarrow h \end{aligned} \qquad (12.17)$$

where $\dot{Q}$ is the heat flux [W] (not to be confused with the volumetric flow rate Q). The definition of the transport coefficient leads to

$$\boxed{\mathrm{d}\dot{Q} = h \, (T_0 - T_\infty) \, \mathrm{d}A} \qquad (12.18)$$

This expression can be helpful to derive the SI units of the heat transport coefficient,

$$[h] = \mathrm{W}/(\mathrm{m}^2 \, \mathrm{K}) \qquad (12.19)$$

An expression for the heat transfer coefficient can be obtained equaling the heat flux calculated from Fourier's law and from the transport coefficient definition,

$$\mathrm{d}\dot{Q} = -\kappa \left. \frac{\partial T}{\partial y} \right|_{y=0} \mathrm{d}A = h \, (T_0 - T_\infty) \, \mathrm{d}A \qquad (12.20)$$

which implies,

$$h = \frac{\kappa \left. \dfrac{\partial T}{\partial y} \right|_{y=0}}{T_\infty - T_0} \qquad (12.21)$$

This expression gives us the order of magnitude of h as a function of κ and the thermal boundary layer thickness δ_T,

$$h = \mathcal{O}\left(\frac{\kappa\dfrac{T_\infty - T_0}{\delta_T}}{T_\infty - T_0}\right) = \mathcal{O}\left(\frac{\kappa}{\delta_T}\right) \tag{12.22}$$

where the symbol $\mathcal{O}(\cdot)$ denotes *order of magnitude* of $(\cdot)$.

Finally, the global transport coefficient is related to the local coefficient as,

$$\overline{h} = \frac{1}{A}\int_A h \; \mathrm{d}A \tag{12.23}$$

and

$$\boxed{\dot{Q} = \overline{h}\;(T_0 - T_\infty)\;A} \tag{12.24}$$

Next, we are going to analyze the two main cases of heat transport: forced convection and natural convection. Between these two cases we may find mixed convection.

12.3.1 Heat Transfer by Forced Convection

Heat transport is said to be by forced convection when it happens in the presence of a free-stream velocity field. One of the traits of forced convection is that the fluid velocity enhances heat transport and, in general, all transport processes.

For the analysis, let us assume that the heat transport coefficient h about an interface is desired in a uniform stream of velocity V. Typically, this parameter is measured experimentally, so let's apply dimensional analysis.

For a steady, incompressible flow, the heat transfer coefficient depends on the free-stream velocity V; the dynamic properties of the fluid ρ, μ; the thermal fluid properties c_p, κ; the geometry and the dimensions of the body or interface, represented by the characteristic length L (here we could include every length and angle of the body, but this is typically implicitly assumed). Thus,

$$h = \varphi(\rho, \mu, L, V, c_p, \kappa) \tag{12.25}$$

The dimensional matrix can be written as

	M	L	T	Θ
ρ	1	-3	0	0
V	0	1	-1	0
L	0	1	0	0
c_p	0	2	-2	-1
κ	1	1	-3	-1
μ	1	-1	-1	0
h	1	0	-3	-1

whose range is 4. The first four variables ρ, V, L, c_p are chosen as dimensionally independent variables. The dimensionless groups are as follows.

- Π_κ
  The requirement that the ratio

$$\Pi_\kappa = \frac{\kappa}{\rho^\alpha V^\beta L^\gamma c_p^\delta}$$

be dimensionless implies

$$\mathsf{M}^1\mathsf{L}^1\mathsf{T}^{-3}\Theta^{-1} = (\mathsf{ML}^{-3})^\alpha\,(\mathsf{LT}^{-1})^\beta\,\mathsf{L}^\gamma\,(\mathsf{L}^2\mathsf{T}^{-2}\Theta^{-1})^\delta$$

$$\begin{cases} 1 & = \alpha \\ 1 & = -3\alpha + \beta + \gamma + 2\delta \\ -3 & = -\beta - 2\delta \\ -1 & = -\delta \end{cases}$$

Finally

$$\Pi_\kappa = \frac{\kappa}{\rho c_p V L} = \frac{1}{\mathrm{Pe}}$$

- Π_μ
  For μ, one obtains the Reynolds number

$$\Pi_\mu = \frac{\mu}{\rho V L} = \frac{1}{\mathrm{Re}}$$

- Π_h
  Similarly,

$$\mathsf{M}^1\mathsf{L}^0\mathsf{T}^{-3}\Theta^{-1} = (\mathsf{ML}^{-3})^\alpha\,(\mathsf{LT}^{-1})^\beta\,\mathsf{L}^\gamma\,(\mathsf{L}^2\mathsf{T}^{-2}\Theta^{-1})^\delta$$

$$\begin{cases} 1 & = \alpha \\ 0 & = -3\alpha + \beta + \gamma + 2\delta \\ -3 & = -\beta - 2\delta \\ -1 & = -\delta \end{cases}$$

from where Π_h equals

$$\boxed{\mathrm{St} = \frac{h}{\rho V c_p} \qquad \text{Stanton number}} \qquad (12.26)$$

The Stanton number is an important dimensionless number in heat transport phenomena.

Thus, the dimensionless expression for the heat transport coefficient can be written as

$$\text{St} = \widetilde{\varphi}(\text{Re}, \text{Pe})$$

This expression can also be found as a function of the Nusselt number,

$$\text{Nu} = \frac{hL}{\kappa} \qquad \text{Nusselt number} \tag{12.27}$$

which can be related to the Stanton number,

$$\text{St} = \frac{h}{\rho V c_p} = \frac{hL}{\kappa}\frac{\kappa}{\rho V c_p L} = \text{Nu}\frac{1}{\text{Pe}} \tag{12.28}$$

Since

$$\text{Pe} = \text{Re Pr}$$

the nondimensional dependency can be rewritten as

$$\text{Nu} = \widetilde{\varphi}'(\text{Re}, \text{Pr}) \tag{12.29}$$

Recall that Nu also represents the dimensionless heat flux at the interface (see Chapter 9).

Remark 12.4. Correlations for the heat transport coefficient typically appear in the form

$$\text{Nu} = C \text{ Re}^a \text{ Pr}^b$$

where C, a and b are constants.

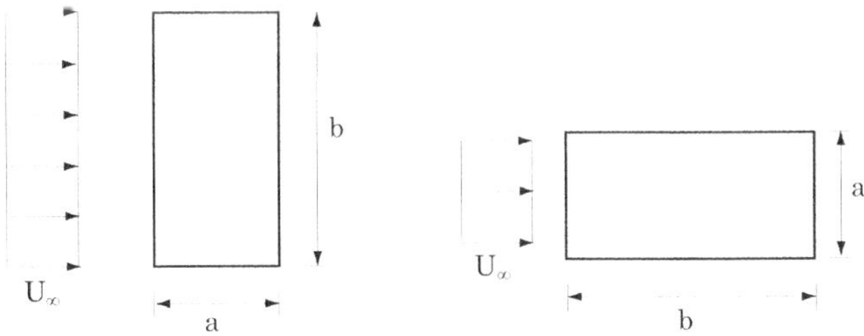

Fig. 12.5. Cooling down a flat plate with a uniform stream.

Example 12.2. It is desired to cool down a thin flat plate of dimensions $a \times b$ with $a << b$ (see Fig. 12.5). The plate rests horizontally on the ground. When a uniform air stream is blown onto the plate, for which orientation will the plate cool down faster?

Solution. Let us assume the conditions are those of laminar flow. The answer to the problem would be the same for turbulent flow conditions. For air (Pr = 0.72) at laminar flow conditions, the global Nusselt number for a flat plate is given by

$$\mathrm{Nu}_L = 0.664 \ \mathrm{Re}_L^{1/2}\mathrm{Pr}^{1/3}$$

where L is the length of the boundary layer along the streamline direction. The total heat flux on the plate is given by

$$\dot{Q} = \bar{h} \ \Delta T \ S$$

with $S = ab$, the surface of the plate. Therefore, for a given temperature difference ΔT the configuration with the highest $\bar{h}$ will cool down faster. Thus,

$$\begin{aligned}
\bar{h} \ &= \frac{\kappa \ \mathrm{Nu}_L}{L} \\
&= 0.664 \ \frac{\kappa}{L} \ \mathrm{Re}_L^{1/2}\mathrm{Pr}^{1/3} \\
&= 0.664 \ \kappa \left(\frac{\rho V}{\mu}\right)^{1/2} \mathrm{Pr}^{1/3} L^{-1/2}
\end{aligned}$$

Note that all the parameters are the same for both configurations except L. The shorter the L, the larger $\bar{h}$. Therefore, we want to pick $L = a$ so the smaller side is oriented in the direction of the flow.

This is so because at the leading edge, the heat transfer rate is largest and the above plate orientation posseses the longest leading edge.

12.3.2 Heat Transfer by Natural Convection

In this case, the flow around the interface is not forced, but is generated by buoyancy forces. These are due to the density differences within the fluid, caused by the thermal gradients between the interface and the free-stream. This type of heat transport is called *natural convection* or *free convection*.

In natural convection, the heat transport cannot depend on the velocity of the forced free-stream V. Rather, it depends on the dynamic fluid properties, ρ, μ; its thermal properties, c_p and κ; the problem characteristic length L; and the characteristic buoyant force, a function of the product $g\beta\Delta T$, where g is the acceleration of the gravity, β the thermal dilatation coefficient (units $1/\mathrm{K}$), and ΔT the characteristic temperature difference of the flow. Summarizing,

$$h = \varphi(\rho, \mu, L, c_p, \kappa, g\beta\Delta T) \tag{12.30}$$

The dimensional matrix is given by

	M	L	T	Θ
ρ	1	-3	0	0
L	0	1	0	0
μ	1	-1	-1	0
c_p	0	2	-2	-1
κ	1	1	-3	-1
$g\beta\Delta T$	0	1	-2	0
h	1	0	-3	-1

whose range is 4. The dimensionally independent variables are chosen as ρ, L, μ and c_p. Next, the dimensionless groups are determined.

- Π_κ

  Requiring the ratio below be nondimensional,

$$\Pi\kappa = \frac{\kappa}{\rho^\alpha L^\beta \mu^\gamma c_p^\delta}$$

  implies

$$M^1 L^1 T^{-3} \Theta^{-1} = (ML^{-3})^\alpha \; L^\beta \; (ML^{-1}T^{-1})^\gamma \; (L^2 T^{-2}\Theta^{-1})^\delta$$

$$\begin{cases} 1 & = \alpha + \gamma \\ 1 & = -3\alpha + \beta - \gamma + 2\delta \\ -3 & = -\gamma - 2\delta \\ -1 & = -\delta \end{cases}$$

  Thus,

$$\Pi_\kappa = \frac{\kappa}{\mu c_p} = \frac{1}{\mathrm{Pr}}$$

- Π_β

  From

$$\Pi_\beta = \frac{g\beta\Delta T}{\rho^\alpha L^\beta \mu^\gamma c_p^\delta}$$

  one gets

$$M^0 L^1 T^{-2} \Theta^0 = (ML^{-3})^\alpha \; L^\beta \; (ML^{-1}T^{-1})^\gamma \; (L^2 T^{-2}\Theta^{-1})^\delta$$

$$\begin{cases} 0 & = \alpha + \gamma \\ 1 & = -3\alpha + \beta - \gamma + 2\delta \\ -2 & = -\gamma - 2\delta \\ 0 & = -\delta \end{cases}$$

Thus Π_β can be written as

$$\boxed{\text{Gr} = \frac{\rho^2 g\beta\,\Delta T\,L^3}{\mu^2} \qquad \text{Grashof number}} \tag{12.31}$$

The Grashof number indicates the relevance of the buoyancy forces due to the thermal gradients compared with the viscous forces. Sometimes, it is replaced by the Rayleigh number, Ra,

$$\boxed{\text{Ra} = \text{Gr Pr} = \frac{\rho g\beta\,\Delta T\,L^3}{\mu\alpha} \qquad \text{Rayleigh number}} \tag{12.32}$$

- Π_h
  Similarly,

$$\mathsf{M}^1\mathsf{L}^0\mathsf{T}^{-3}\Theta^{-1} = (\mathsf{ML}^{-3})^\alpha\ \mathsf{L}^\beta\ (\mathsf{ML}^{-1}\mathsf{T}^{-1})^\gamma\ (\mathsf{L}^2\mathsf{T}^{-2}\Theta^{-1})^\delta$$

$$\begin{cases} 1 & = \alpha + \gamma \\ 0 & = -3\alpha + \beta - \gamma + 2\delta \\ -3 & = -\gamma - 2\delta \\ -1 & = -\delta \end{cases}$$

$$\Pi_h = \frac{hL}{\mu c_p}$$

The dimensionless dependency of the heat transport coefficient can be finally written as

$$\boxed{\frac{hL}{\mu c_p} = \widetilde{\varphi}(\text{Pr}, \text{Gr})} \tag{12.33}$$

The nondimensional heat transport coefficient can also be expressed as a function of the Nusselt number,

$$\frac{hL}{\mu c_p} = \frac{hL}{\kappa}\,\frac{\kappa}{\mu c_p} = \text{Nu}\frac{1}{\text{Pr}} \tag{12.34}$$

Thus, for natural convection, the dimensionless heat transport coefficient is a function of Prandtl and Grashof:

$$\boxed{\text{Nu} = \widetilde{\varphi}'(\text{Pr}, \text{Gr})} \tag{12.35}$$

or Prandtl and Rayleigh

$$\boxed{\text{Nu} = \widetilde{\varphi}''(\text{Pr}, \text{Ra})} \tag{12.36}$$

12.3.3 Basic Heat Transport Coefficients

The following correlations are extracted from [16, 10, 3] and are for constant property boundary layers. This assumption is admissible for many heat transport problems.

Table 12.5. Basic heat transport coefficients. Forced convection.

Forced convection		
Geometry	Regime	Correlation
flat plate	laminar, isothermal $\mathrm{Pr} > 0.6$ $\mathrm{Re}_x < 3 \times 10^5$	$\mathrm{Nu}_x = 0.332\ \mathrm{Re}_x^{1/2}\mathrm{Pr}^{1/3}$
flat plate	laminar, isothermal $\mathrm{Pr} > 0.6$ $\mathrm{Re}_x < 3 \times 10^5$	$\mathrm{Nu}_L = 0.664\ \mathrm{Re}_L^{1/2}\mathrm{Pr}^{1/3}$
flat plate	laminar, isothermal $\mathrm{Pr} \ll 1$ $\mathrm{Re}_x < 3 \times 10^5$	$\mathrm{Nu}_x = 0.565\ \mathrm{Re}_x^{1/2}\mathrm{Pr}^{1/2}$
flat plate	laminar, constant heat-flux $\mathrm{Re}_x < 3 \times 10^5$	$\mathrm{Nu}_x = 0.453\ \mathrm{Re}_x^{1/2}\mathrm{Pr}^{1/3}$
flat plate	turbulent, isothermal $0.5 < \mathrm{Pr} < 1$ $5 \times 10^5 < \mathrm{Re}_x < 5 \times 10^6$	$\mathrm{Nu}_x = 0.0287\ \mathrm{Re}_x^{4/5}\mathrm{Pr}^{3/5}$
flat plate	turbulent, isothermal $0.6 < \mathrm{Pr} < 60$ $5 \times 10^5 < \mathrm{Re}_x < 5 \times 10^8$	$\mathrm{Nu}_L = 0.037\ (\mathrm{Re}_L^{4/5} - 871)\ \mathrm{Pr}^{1/3}$
flat plate	turbulent, isothermal	$\mathrm{Nu}_L = 0.360\ \mathrm{Re}_L^{4/5}\mathrm{Pr}^{1/3}$
pipe	laminar, isothermal, fully developed T and u	$\mathrm{Nu}_D = 3.658$
pipe	laminar, constant flux, fully developed T and u	$\mathrm{Nu}_D = 4.364$
cylinder	isothermal	$\mathrm{Nu} = (0.4\ \mathrm{Re}_D^{1/2} + 0.06\ \mathrm{Re}_D^{2/3})$ $\times\ \mathrm{Pr}^{2/5}(\frac{\mu_\infty}{\mu_0})^{1/4}$
sphere	isothermal	$\mathrm{Nu}_D = 2 + 0.6\ \mathrm{Re}_D^{1/2}\mathrm{Pr}^{1/3}$
packed bed of spheres	isothermal d = spheres diameter $\rho V = \dot{m}/A$ A = cross-sectional area	$\mathrm{Nu}_d = 1.625\ \mathrm{Re}_d^{1/2}\mathrm{Pr}^{1/3}$

Table 12.6. Basic heat transport coefficients. Natural convection.

Natural convection

Geometry	Regime	Correlation
vertical plate	laminar, isothermal[†]	$\mathrm{Nu}_x = f(\mathrm{Pr})\,\mathrm{Gr}_x^{1/4}$
vertical plate	laminar, isothermal[†]	$\mathrm{Nu}_L = \frac{4}{3} f(\mathrm{Pr})\,\mathrm{Gr}_L^{1/4}$
vertical plate	laminar, isothermal	$\mathrm{Nu}_x = 0.508 \left(\frac{\mathrm{Pr}^2}{0.952+\mathrm{Pr}} \right)^{1/4} \mathrm{Gr}_x^{1/4}$
vertical plate	turbulent, isothermal $10^9 < \mathrm{Gr}_L < 10^{12}$	$\mathrm{Nu}_L = \frac{0.15\ (\mathrm{Gr}_L\mathrm{Pr})^{1/3}}{[1+(0.492/\mathrm{Pr})^{9/16}]^{16/27}}$
vertical plate	turbulent, constant heat-flux $10^9 < \mathrm{Gr}_L$	$\mathrm{Nu}_x = 0.0942\ (\mathrm{Gr}_x\mathrm{Pr})^{1/3}$

[†]See Table 12.7 for $f(\mathrm{Pr})$.

Table 12.7. Function $f(\mathrm{Pr})$ for the heat transfer coefficient Nu on an isothermal vertical surface due to natural convection [16].

Pr	0.01	0.1	0.72	1.0	10	100	1000
$f(\mathrm{Pr})$	0.0570	0.164	0.357	0.401	0.827	1.55	2.80

12.4 Mass Transport

The formulation of the mass transport coefficient, h_m or K_m, is analogous to that of the heat transfer coefficient. In this case, we identify

$$\Gamma_0 \longrightarrow J_A$$
$$\Pi \longrightarrow \rho_A \tag{12.37}$$
$$K \longrightarrow h_m$$

where J_A is the mass flux (kg/s) of the chemical species A. The definition of the mass transport coefficient is

$$\boxed{\mathrm{d}J_A = h_m \left(\rho_{A0} - \rho_{A\infty} \right) \mathrm{d}A} \tag{12.38}$$

The SI units of the mass transport coefficient can be deduced to be

$$[h_m] = \frac{\mathrm{m}}{\mathrm{s}} \tag{12.39}$$

Assuming small variations of mixture density, through the Fick's law the mass transfer coefficient can be related to the derivative of the concentration at the interface,

$$\mathrm{d}J_A = -D_{AB} \left.\frac{\partial \rho_A}{\partial y}\right|_{y=0} \mathrm{d}A = h_m \left(\rho_{A0} - \rho_{A\infty}\right) \mathrm{d}A \tag{12.40}$$

from where

$$h_m = \frac{D_{AB}\left.\dfrac{\partial \rho_A}{\partial y}\right|_{y=0}}{\rho_{A\infty} - \rho_{A0}} \tag{12.41}$$

This equation shows that the order of magnitude, denoted by $\mathcal{O}(\cdot)$, of the mass transport coefficient is related to the molecular diffusion D_{AB} and the mass boundary layer thickness δ_c,

$$h_m = \mathcal{O}\left(\frac{D_{AB}\dfrac{\rho_{A\infty} - \rho_{A0}}{\delta_c}}{\rho_{A\infty} - \rho_{A0}}\right) = \mathcal{O}\left(\frac{D_{AB}}{\delta_c}\right) \tag{12.42}$$

Finally, the global mass transfer coefficient can be related to the average of the local coefficient,

$$\overline{h}_m = \frac{1}{A}\int_A h_m \, \mathrm{d}A \tag{12.43}$$

for which

$$\boxed{J_A = \overline{h}_m \left(\rho_{A0} - \rho_{A\infty}\right) A} \tag{12.44}$$

Other Mass Transfer Coefficients

The definition of transport coefficient can be applied to other measures of concentration, such as mass fraction or molar concentration.

As a function of mass fractions, equation (12.38) can be written as

$$\boxed{\mathrm{d}J_A = h_Y \left(Y_{A0} - Y_{A\infty}\right) \mathrm{d}A} \tag{12.45}$$

from where

$$h_Y = \rho h_m \tag{12.46}$$

Similarly, a transport coefficient can be defined for the molar flux J_A^m as a function of the molar concentration or molar fraction,

$$\begin{aligned}
\mathrm{d}J_A^m &= h_c \left(c_{A0} - c_{A\infty}\right) \mathrm{d}A \tag{12.47}\\
&= c h_X \left(X_{A0} - X_{A\infty}\right) \mathrm{d}A \tag{12.48}
\end{aligned}$$

from where it can be concluded

$$h_X = c \, h_c \tag{12.49}$$

According to Dalton's law, the molar fraction of a component in a gas X_A can be written as a function of the partial pressure p_A. Thus, using the total pressure of the mixture p,

$$\begin{aligned} \mathrm{d}J_A^m &= h_X\,(X_{A0} - X_{A\infty})\,\mathrm{d}A \\ &= h_X\frac{1}{p}\,(p_{A0} - p_{A\infty})\,\mathrm{d}A \\ &= h_p\,(p_{A0} - p_{A\infty})\,\mathrm{d}A \end{aligned} \qquad (12.50)$$

where the corresponding mass transfer coefficient is in this case

$$h_p = \frac{h_X}{p} = \frac{c\,h_c}{p} = \frac{h_c}{\mathcal{R}T} \qquad (12.51)$$

where $\mathcal{R}$ is the universal gas constant.

12.4.1 Mass Transport by Forced Convection

To study the mass transport coefficient by forced convection, we proceed in a similar fashion to previous sections. The mass transfer coefficient depends on the free-stream velocity V; the fluid dynamic properties ρ, μ; the mass diffusion coefficient D_{AB}; and a characteristic length scale of the problem L. Thus,

$$h_m = \varphi(\rho, \mu, L, V, D_{AB}) \qquad (12.52)$$

This yields the following dimensional matrix,

	M	L	T
ρ	1	-3	0
V	0	1	-1
L	0	1	0
μ	1	-1	-1
D_{AB}	0	2	-1
h_m	0	1	-1

whose range is 3. The first three variables, ρ, V, L, are chosen as the dimensionally independent variables, leading to the following dimensionless variables,

$$\begin{aligned} \Pi_\mu &= \frac{\mu}{\rho V L} = \frac{1}{\mathrm{Re}} \\ \Pi_{D_{AB}} &= \frac{D_{AB}}{V L} = \frac{1}{\mathrm{Pe_{II}}} \\ \Pi_{h_m} &= \frac{h_m}{V} = \mathrm{St_m} \end{aligned}$$

The last number is called the *mass transfer Stanton number*,

$$\boxed{\text{St}_\text{m} = \frac{h_m}{V} \qquad \text{Mass transfer Stanton number}} \tag{12.53}$$

Finally, the nondimensional dependency of the mass transfer coefficient is

$$\boxed{\text{St}_\text{m} = \widetilde{\varphi}(\text{Re}, \text{Pe}_\text{II})} \tag{12.54}$$

Likewise, the nondimensional mass transport coefficient can be expressed as a *mass Nusselt or Sherwood number*,

$$\boxed{\text{Nu}_m = \text{Sh} = \frac{h_m L}{D_{AB}} \qquad \text{Mass Nusselt or Sherwood number}} \tag{12.55}$$

Thus,

$$\text{St}_\text{m} = \frac{h_m}{V} = \frac{h_m L}{D_{AB}} \frac{D_{AB}}{V L} = \text{Nu}_m \frac{1}{\text{Pe}_\text{II}} \tag{12.56}$$

Furthermore,

$$\text{Pe}_\text{II} = \text{Re}\,\text{Sc} \tag{12.57}$$

and another way to write out the nondimensional dependency is

$$\boxed{\text{Nu}_m = \text{Sh} = \widetilde{\varphi}'(\text{Re}, \text{Sc})} \tag{12.58}$$

12.4.2 Mass Transport by Natural Convection

As for heat transport, it is also possible to transfer mass induced by density variations in a gravity field. In this case, the mass transfer coefficient depends on the fluid dynamic properties ρ, μ; the mass diffusion coefficient D_{AB}; the characteristic concentration difference $g\Delta\rho_A$; and a characteristic length scale of the problem L. Thus,

$$h_m = \varphi(\rho, \mu, L, g\Delta\rho_A, D_{AB}) \tag{12.59}$$

In dimensionless form, the functional dependency can be expressed as

$$\boxed{\text{Nu}_m = \text{Sh} = \widetilde{\varphi}(\text{Gr}_\text{m}, \text{Sc})} \tag{12.60}$$

where the concentration Grashof number is

$$\boxed{\text{Gr}_\text{m} = \frac{\rho g \Delta\rho_A L^3}{\mu^2} \qquad \text{concentration Grashof number}} \tag{12.61}$$

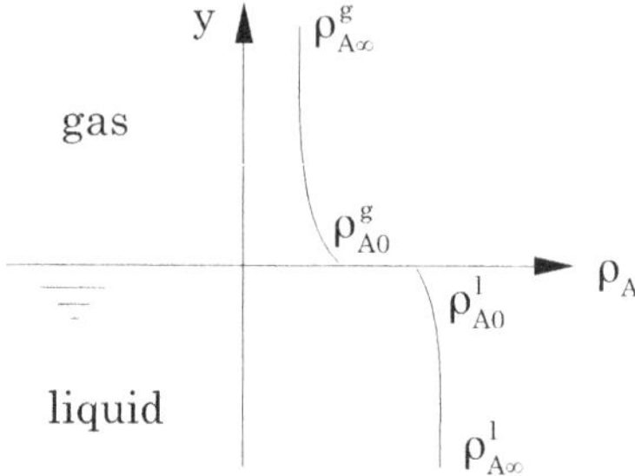

Fig. 12.6. Concentration profile of species A across an interface.

12.4.3 Mass Transfer across Fluid/Fluid Interfaces

Assume that there is a liquid/gas interface, like that of Fig. 12.6. The same procedure can be applied to a general fluid/fluid interface.

At each side of the interface there is a distinct mass transfer coefficient. In the absence of surface chemical reactions and considering that the interface cannot store mass, the mass flux of species A across the interface is continuous,

$$j_A^g = j_A^l \tag{12.62}$$

Here the superscripts g and l refer to the gas and liquid phase, respectively. Applying the definition of the mass transport coefficient at both sides of the interface,

$$j_A = h_m^g(\rho_{A0}^g - \rho_{A\infty}^g) = h_m^l(\rho_{A\infty}^l - \rho_{A0}^l) \tag{12.63}$$

where the flux has been taken positive from the liquid towards the gas.

In general, the chemical concentrations are not continuous across the interface, and so, $\rho_{A0}^g \neq \rho_{A0}^l$, but these are related by the equilibrium curve and solubility data,

$$\rho_{A0}^g = f(\rho_{A0}^l) \tag{12.64}$$

As exceptions where this is not a good approximation we can cite extremely high reaction rates (typical in near vacuum conditions) and interfaces with high concentrations of particles or surfactants.

Example 12.3 (Evaporation across a water/air interface). The water evaporation rate is to be obtained for the pool of Fig. 12.7 when the wind blows from the West at 1 m/s. The size of the pool is given by $a = 9$ m and $b = 6$ m. The ambient conditions are $T = 300$ K and 40% relative humidity. At these conditions, the kinematic viscosity of air is $\nu = 1.57 \times 10^{-5}$ m^2/s and the molecular diffusivity of water vapor in air, $D = 0.26 \times 10^{-4}$ m^2/s.
Solution. The Reynolds number can be calculated as $\mathrm{Re}_L = 2.5 \times 10^5$ so assuming that the boundary layer starts at the pool, the laminar correlation can be used,

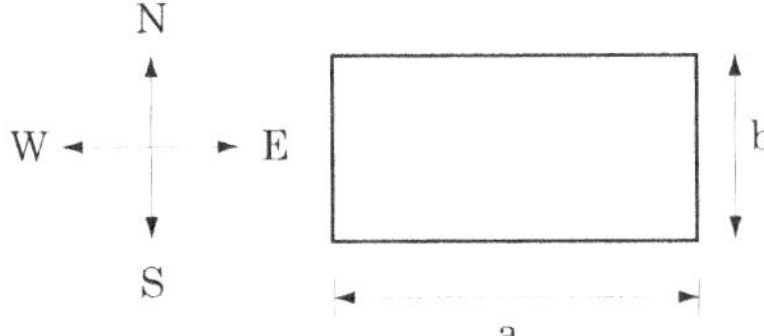

Fig. 12.7. Evaporation across a water/air interface.

$$\mathrm{Sh}_L = 0.664\ \mathrm{Re}_L^{1/2}\mathrm{Sc}^{1/3}$$

where for air, $\mathrm{Sc} = 0.6$. Thus, $\mathrm{Sh}_L = 280.0$ and $\bar{h}_m = \mathrm{Sh}_L D/L = 8.1 \times 10^{-4}$ m/s.

The mass flux can be calculated from the global transport coefficient,

$$J_{\mathrm{H_2O}} = \bar{h}_m \left(\rho_{\mathrm{H_2O}_0} - \rho_{\mathrm{H_2O}_\infty}\right)A$$

The concentration at infinity can be evaluated from the relative humidity $\phi = 40/100 = 0.4$ and the saturation concentration,

$$\rho_{\mathrm{H_2O}_\infty} = \phi\ \rho_{\mathrm{H_2O}_{\mathrm{sat}}}$$

Then, since the evaporation process is not very fast, we can assume that at the interface the water vapor is at equilibrium conditions, $\rho_{\mathrm{H_2O}_0} = \rho_{\mathrm{H_2O}_{\mathrm{sat}}}$. Thus,

$$J_{\mathrm{H_2O}} = \bar{h}_m \left(1 - \phi\right)\rho_{\mathrm{H_2O}_{\mathrm{sat}}} A$$

The equilibrium conditions at the interface are obtained from the water-air saturation curve at 300 K, $p_{\mathrm{sat}} = 3\,531$ Pa, and because $M_{\mathrm{H_2O}} = 18$

$$\rho_{\mathrm{H_2O}_{\mathrm{sat}}} = \frac{p_{\mathrm{sat}}}{R_{\mathrm{gas}}T} = \frac{3\,531}{\dfrac{8\,314}{18}\,300} = 0.025\ \mathrm{kg/m^3}$$

Gathering everything, one obtains,

$$J_{\mathrm{H_2O}} = 1.5 \times 10^{-4}\ \mathrm{kg/s} = 13\ \mathrm{kg/day}$$

Overall Mass Transfer Coefficient

Sometimes, in order to avoid calculating the equilibrium concentrations at the interface, we may want to obtain the mass transfer rate from the bulk concentrations in the fluids.

Let us define the equilibrium concentrations, ρ_{Ae}^l and ρ_{Ae}^g which correspond, respectively, to $\rho_{A\infty}^g$ and $\rho_{A\infty}^l$, that is

$$\rho_{A\infty}^{g} \;=\; f(\rho_{Ae}^{l}) \tag{12.65}$$

$$\rho_{Ae}^{g} \;=\; f(\rho_{A\infty}^{l}) \tag{12.66}$$

Then, the overall mass transfer coefficients K_m^l, K_m^g are defined such that the same mass transfer rate is obtained from bulk concentrations,

$$h_m^g(\rho_{A0}^{g} - \rho_{A\infty}^{g}) \;=\; K_m^g(\rho_{Ae}^{g} - \rho_{A\infty}^{g}) \tag{12.67}$$

$$h_m^l(\rho_{A\infty}^{l} - \rho_{A0}^{l}) \;=\; K_m^l(\rho_{A\infty}^{l} - \rho_{Ae}^{l}) \tag{12.68}$$

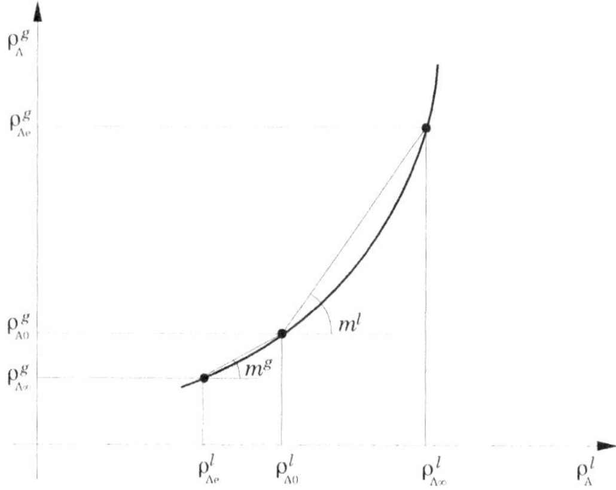

Fig. 12.8. Reference values at the equilibrium phase curve to calculate the overall mass transfer coefficient across an interface.

Defining the slopes at the equilibrium curve,

$$m^l \;=\; \frac{\rho_{Ae}^{g} - \rho_{A0}^{g}}{\rho_{A\infty}^{l} - \rho_{A0}^{l}} \tag{12.69}$$

$$m^g \;=\; \frac{\rho_{A0}^{g} - \rho_{A\infty}^{g}}{\rho_{A0}^{l} - \rho_{Ae}^{l}} \tag{12.70}$$

it can be concluded that

$$\frac{1}{K_m^l} \;=\; \frac{1}{h_m^l} + \frac{1}{m^l h_m^g} \tag{12.71}$$

$$\frac{1}{K_m^g} \;=\; \frac{1}{h_m^g} + \frac{m^g}{h_m^l} \tag{12.72}$$

Three cases can be distinguished.

(a) $\frac{h_m^l}{m^l h_m^g} << 1$. The liquid phase has more resistance than the gas phase and it is said that the mass transfer at the interface is liquid-phase controlled. Then $K_m^l \approx h_m^l$.

(b) $\frac{h_m^l}{m^l h_m^g} >> 1$. The gas phase has more resistance than the liquid phase and it is said that the mass transfer at the interface is gas-phase controlled. Then $K_m^g \approx h_m^g$.

(c) $0.01 < \frac{h_m^l}{m^l h_m^g} < 10$. Then both phases present similar resistance to mass transfer and the interactions between the phases must be accounted for.

12.4.4 Basic Mass Transport Coefficients

The following correlations are extracted from [16, 10, 3] and are for constant property boundary layers. This assumption may be a rude approximation for mass transfer problems, mainly when mass transfer rates are high.

Table 12.8. Basic mass transport coefficients.

Forced convection		
Geometry	Regime	Correlation
flat plate	laminar, iso-concentration $Sc > 0.6$ $Re < 3 \times 10^5$	$Sh_x = 0.332 \, Re_x^{1/2} Sc^{1/3}$
flat plate	laminar, iso-concentration $Sc > 0.6$ $Re < 3 \times 10^5$	$Sh_L = 0.664 \, Re_L^{1/2} Sc^{1/3}$
flat plate	laminar, constant mass flux $Re_x < 3 \times 10^5$	$Sh_x = 0.453 \, Re_x^{1/2} Sc^{1/3}$
sphere	iso-concentration	$Sh_D = 2 + 0.6 \, Re_D^{1/2} Sc^{1/3}$

12.5 Analogies

In previous chapters, we have seen that there is a strong similitude among the mathematical models of transport phenomena for both convection and diffusion. Transport by convection is modeled by the scalar product of the velocity times a spatial gradient. Chapter 7 showed that the constitutive relations are similar: the diffusion flux equals a diffusion coefficient times a gradient.

These similarities are not by chance, but are due to the physical transport mechanisms being the same for all the fluid variables. Furthermore, these similarities can be mathematically explained and lead to the so-called *analogies* between the various transport phenomena.

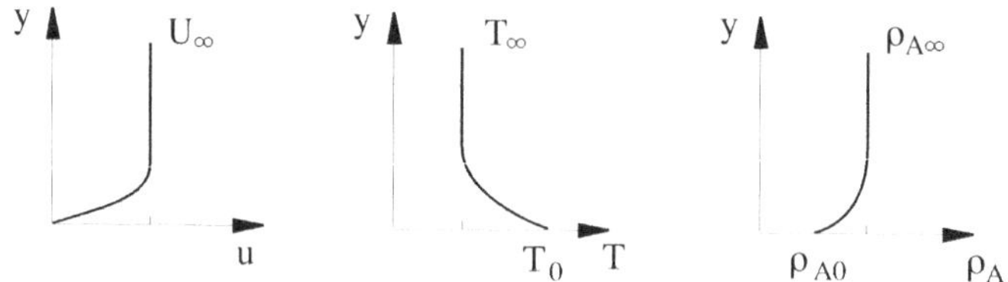

Fig. 12.9. Behavior of the dimensional variables in the neighborhood of a solid boundary.

12.5.1 Reynolds Analogy

Consider the dimensional velocity, temperature and chemical concentration around a solid boundary, as Fig. 12.9 depicts. The subscript 0 denotes the values at the boundary, whereas the infinity, the free-stream values.

Let us define the following dimensionless variables,

$$
\begin{aligned}
x^* &= \frac{x}{L} & u^* &= \frac{u}{U_\infty} \\
T^* &= \frac{T - T_0}{T_\infty - T_0} & \rho_A^* &= \frac{\rho_A - \rho_{A0}}{\rho_{A\infty} - \rho_{A0}} \\
p^* &= \frac{p}{\rho U^2}
\end{aligned}
\tag{12.73}
$$

With this change of variables, all the dimensionless variables obey the same

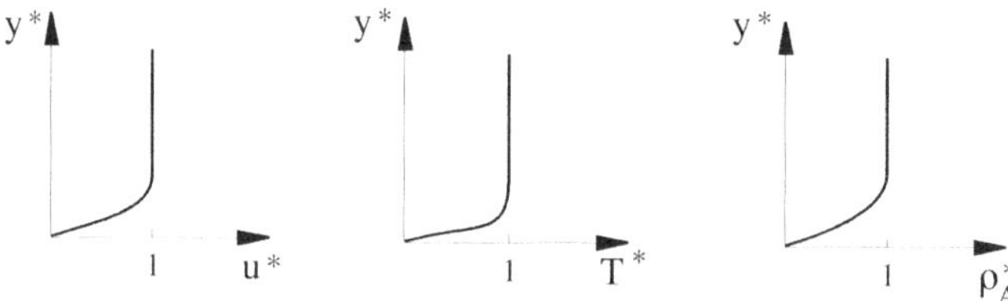

Fig. 12.10. Behavior of the dimensionless variables in the neighborhood of the solid boundary.

boundary conditions across the boundary layer. At the solid boundary, $y = 0$, $u = 0$, $T = T_0$, $\rho_A = \rho_{A0}$ and, therefore,

$$
y = 0 \qquad
\begin{cases}
u^* &= 0 \\
T^* &= 0 \\
\rho_A^* &= 0
\end{cases}
\tag{12.74}
$$

Away from the solid boundary, at $y = \infty$, $\boldsymbol{u} = \boldsymbol{U}_\infty$, $T = T_\infty$ and $\rho_A = \rho_{A\infty}$ and, therefore,

$$y \to \infty \quad \begin{cases} u^* = 1 \\ T^* = 1 \\ \rho_A^* = 1 \end{cases} \tag{12.75}$$

See Fig. 12.10.

We introduce the above change of variables in the two-dimensional, steady, constant fluid property flow equations, with negligible viscous dissipation and no chemical reactions. The dimensionless form of the equations is

$$\boldsymbol{u}^* \cdot \nabla \boldsymbol{u}^* = -\nabla p^* + \frac{1}{\mathrm{Re}} \nabla^2 \boldsymbol{u}^* \tag{12.76}$$

$$\boldsymbol{u}^* \cdot \nabla T^* = + \frac{1}{\mathrm{Re}\,\mathrm{Pr}} \nabla^2 T^* \tag{12.77}$$

$$\boldsymbol{u}^* \cdot \nabla \rho_A^* = + \frac{1}{\mathrm{Re}\,\mathrm{Sc}} \nabla^2 \rho_A^* \tag{12.78}$$

Note that if $\mathrm{Pr} = \mathrm{Sc}$, Eq. (12.77) and (12.78) are the same. Furthermore, in the absence of pressure gradients, $\nabla p^* = 0$ and for $\mathrm{Pr} = \mathrm{Sc} = 1$, the three equations are equal.

Let us calculate the nondimensional form of the transport coefficients. All the derivatives are calculated at the wall, $y = y^* = 0$.

(a) Momentum

$$C_\mathrm{f} = \frac{\tau_0}{\frac{1}{2}\rho U_\infty^2} = \frac{\mu \frac{\partial u}{\partial y}\big|_{y=0}}{\frac{1}{2}\rho U_\infty^2} = \frac{\mu \frac{U_\infty}{L} \frac{\partial u^*}{\partial y^*}\big|_{y^*=0}}{\frac{1}{2}\rho U_\infty^2} = \frac{2}{\mathrm{Re}} \frac{\partial u^*}{\partial y^*}\Big|_{y^*=0} \tag{12.79}$$

(b) Heat

$$q = -\kappa \frac{\partial T}{\partial y}\Big|_{y=0} = -\kappa \frac{T_\infty - T_0}{L} \frac{\partial T^*}{\partial y^*}\Big|_{y^*=0} \tag{12.80}$$

$$h(T_0 - T_\infty) = \kappa \frac{T_0 - T_\infty}{L} \frac{\partial T^*}{\partial y^*}\Big|_{y^*=0} \tag{12.81}$$

$$\mathrm{Nu} = \frac{hL}{\kappa} = \frac{\partial T^*}{\partial y^*}\Big|_{y^*=0} \tag{12.82}$$

(c) Mass

$$j_A = -D_{AB} \frac{\partial \rho_A}{\partial y}\Big|_{y=0} = -D_{AB} \frac{\rho_{A\infty} - \rho_{A0}}{L} \frac{\partial \rho_A^*}{\partial y^*}\Big|_{y^*=0} \tag{12.83}$$

$$h_m(\rho_{A0} - \rho_{A\infty}) = D_{AB} \frac{\rho_{A0} - \rho_{A\infty}}{L} \frac{\partial \rho_A^*}{\partial y^*}\Big|_{y^*=0} \tag{12.84}$$

$$\mathrm{Sh} = \frac{h_m L}{D_{AB}} = \frac{\partial \rho_A^*}{\partial y^*}\Big|_{y^*=0} \tag{12.85}$$

That is, the dimensionless transport coefficients are related to the gradients of the nondimensional variables.

In the absence of pressure gradients, $\nabla p^* = 0$ (i.e. at constant pressure) and if $\mathrm{Pr} = \mathrm{Sc} = 1$, then, the three dimensionless equations and boundary conditions are the same. Thus, the dimensionless solutions must be also the same for the three fluid variables:

$$u^*(x^*, y^*) = T^*(x^*, y^*) = \rho_A^*(x^*, y^*) \tag{12.86}$$

Therefore, the dimensionless derivatives are also equal for the three fields,

$$\frac{\partial u^*}{\partial y^*} = \frac{\partial T^*}{\partial y^*} = \frac{\partial \rho_A^*}{\partial y^*} \tag{12.87}$$

and one arrives at the *Reynolds analogy*:

$$\boxed{\frac{\mathrm{Re}}{2}\, C_f = \mathrm{Nu} = \mathrm{Sh}} \tag{12.88}$$

As a function of Stanton numbers,

$$\boxed{\frac{1}{2}\, C_f = \mathrm{St} = \mathrm{St_m}} \tag{12.89}$$

Remark 12.5. The conditions for this analogy to hold are two: $\mathrm{Pr} = \mathrm{Sc} = 1$ and $\nabla p = 0$. This second condition limits the application of the analogy to configurations where all the drag comes from friction, that is, there is no drag due to pressure or *form drag*. Therefore, this holds for transport coefficients in boundary layers along flat walls and pipes. In particular, this analogy is not applicable for transport coefficients around blunt bodies, like spheres.

12.5.2 Chilton-Colburn Analogy

The Reynolds analogy is quite restrictive because it only can be applied for $\mathrm{Pr} = \mathrm{Sc} = 1$. In practice, it is observed that this relation can be applied for a wider range of dimensionless numbers if a correction as a function of Pr and Sc is introduced. In this way, the *Chilton-Colburn analogy* is obtained,

$$\boxed{\begin{array}{lll} \frac{1}{2} C_f &= \mathrm{St}\, \mathrm{Pr}^{2/3} & 0.6 < \mathrm{Pr} < 60 \\[1ex] \frac{1}{2} C_f &= \mathrm{St_m}\, \mathrm{Sc}^{2/3} & 0.6 < \mathrm{Sc} < 6000 \end{array}} \tag{12.90}$$

Note that this analogy is only applicable within the specified range.

Within both ranges of Prandtl and Schmidt numbers, one can conclude

$$\boxed{\mathrm{Sh} = \mathrm{Nu}\, \mathrm{Le}^{1/3} \qquad \begin{cases} 0.6 < \mathrm{Pr} < 60 \\[1ex] 0.6 < \mathrm{Sc} < 6000 \end{cases}} \tag{12.91}$$

Remark 12.6. This analogy extends the applicability of the Reynolds analogy to Prandtl and Schmidt numbers different from 1. But still, the analogy holds for configurations where all the drag comes from friction, that is, there is no *form drag*. Therefore, it holds for transport coefficients in boundary layers along flat walls and pipes. In particular, this analogy is not applicable for transport coefficients around blunt bodies, like spheres, but there are ways around this.

Remark 12.7. Typically, the range of admissible Pr numbers excludes from the analogy heavy oils and liquid metals.

Remark 12.8. The Chilton-Colburn analogy is presented frequently as a function of the so-called *j-factors*,

$$j_H = j_D = \frac{1}{2}C_f \qquad (12.92)$$

where

$$\begin{aligned} j_H &= \text{St Pr}^{2/3} \\ j_D &= \text{St}_m \text{ Sc}^{2/3} \end{aligned} \qquad (12.93)$$

Remark 12.9. Other analogies were proposed by Prandtl and von Karman, but the Chilton-Colburn is preferable from the simplicity standpoint.

Example 12.4 (Chilton-Colburn analogy). The local drag coefficient for a flat plate of length L in laminar flow can be correlated as

$$C_{fx} = 0.664 \text{ Re}_x^{-1/2}$$

Determine the local heat transport coefficient for an isothermal plate.
Solution. Since there is no form drag, the Chilton-Colburn analogy can be applied for $0.6 \leq \text{Pr} \leq 60$,

$$\frac{1}{2} C_f = \text{St Pr}^{2/3}$$

Substituting the correlation for the drag coefficient and taking into account that $\text{St} = \text{Nu}/\text{Pe}$,

$$\text{Nu}_x = 0.332 \text{ Re}_x^{1/2} \text{ Pr}^{1/3}$$

For variable fluid properties, the above formula should be applied at the *film temperature*

$$T_f = \frac{1}{2}(T_0 + T_\infty)$$

Problems

12.1   An oil with a kinematic viscosity of $\nu = \mu/\rho = 10^{-4}$ m^2/s and a density of $\rho = 800$ kg/m^3 flows through a horizontal pipe of $D = 0.1$ m diameter at a volumetric flow rate of $Q = 0.5$ l/s. Calculate the pressure loss in 10 m of length. (Hint: check whether the flow is laminar or turbulent).

12.2   The pressure loss in a pipe is to be determined through experiments with water ($\nu = 10^{-6}$ m^2/s, $\rho = 10^3$ kg/m^3). If the pressure loss is 130 000 Pa for a water flow rate of 15 kg/s, what is the pressure loss for 20 kg/s of liquid oxygen ($\rho = 1121$ kg/m^3)? It will be assumed that the friction factor is the same for both cases.

12.3   A 280 km oil pipeline connects two pumping stations. It is desired to pump 0.56 m^3/s through a 0.62 m diameter pipe to the exit station which is 250 m below the inlet station. The gage pressure at the exit station must be maintained at $p_e = 300\,000$ Pa. Calculate the power required to pump the oil, which has a kinematic viscosity of 4.5×10^{-6} m^2/s and a density of 810 kg/m^3. The friction factor can be taken equal to $\lambda = 0.015$. The inlet pressure can be assumed atmospheric.

12.4   A long 20 mm diameter cylinder of naphthalene, used in mothballs to repel insects, is exposed to an air stream that has a mean mass transport coefficient of $\bar{K}_m = 0.05$ m/s. The vapor concentration of naphthalene at the cylinder surface is 6.4×10^{-4} kg/m^3. How much naphthalene sublimates per unit length of the cylinder?

12.5   The chips of an electronic circuit are cooled down by an air stream of $T_\infty = 25\,°$C and $V = 10$ m/s. One of the chips is a square of 4 mm $\times$ 4 mm and is placed at 120 mm of the leading edge of the electronic board. Experiments have shown that the Nusselt number based on the distance to the leading edge x can be correlated as

$$\mathrm{Nu}_x = 0.04\ \mathrm{Re}_x^{0.85}\ \mathrm{Pr}^{1/3}$$

$$\mathrm{Nu}_x = \frac{h_x x}{\kappa} \qquad \mathrm{Re}_x = \frac{\rho U x}{\mu}$$

Estimate the temperature of the surface of the chip if it dissipates 30 mW. *Data for air:* $\mu = 1.8 \times 10^{-5}$ kg/(m s), $\rho = 1.2$ kg/m^3, $\kappa = 0.026$ W/(m K), Pr $= 0.7$.

12.6   A series of experiments about heat transfer on a flat plate with a very rough surface show that Nu_x could be correlated as

$$\mathrm{Nu}_x = 0.04\ \mathrm{Re}_x^{0.9}\ \mathrm{Pr}^{1/3}$$

Obtain an expression for the ratio between the global $\bar{h}_L$ and local h_x heat transfer coefficients ($\bar{h}_L/h_x$).

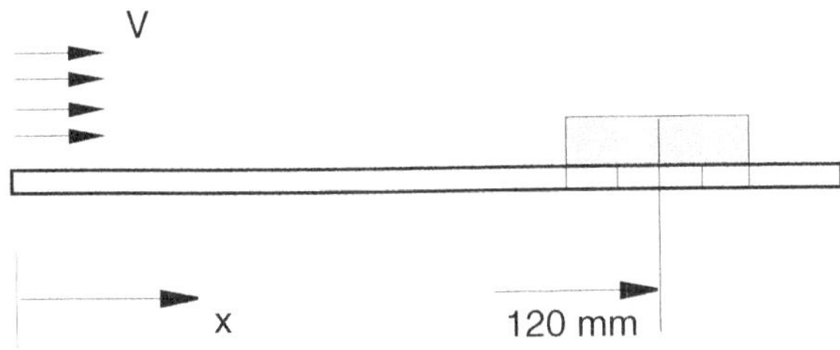

Problem 12.5. Cooling of an electronic chip by forced convection.

12.7   The water evaporation rate of a lake can be calculated by measuring the decrease of the free surface level. Consider a summer day, where the water and ambient air are at 305 K, and the air relative humidity is 40%. If the lake water level decreases at a rate of 0.1 mm/h, how much water is evaporating per unit time and surface? What is the mass transfer coefficient?
Saturation pressure at 305 K: $P_s = 3531$ Pa.

12.8   Photosynthesis, which takes place at the leaves and green areas of the plants, produces a transport of carbon dioxide (CO_2) from the atmosphere to the chloroplasts of the leaves. Therefore, the speed of photosynthesis can be quantified as a function of the assimilation rate of CO_2 by a leaf, which is strongly influenced by the concentration boundary layer about the leaf. If the density of CO_2 in the air and at the leaf surface is, respectively, 6×10^{-4} kg/m^3 and 5×10^{-4} kg/m^3, and the mass transfer coefficient around a leaf is 10^{-2} m/s, calculate the rate of assimilation of CO_2 per unit time and surface of the leaf.

12.9   Chemical species A evaporates from a plane surface to the species B. The concentration profile of A in the boundary layer can be approximated by $C_A(y) = Dy^2 + Ey + F$ with D, E and F constants for any position x. The coordinate y is normal to the surface. Calculate the mass transfer coefficient K_m as a function of the above constants, the concentration of A in fluid B $C_{A\infty}$, and the mass diffusion coefficient D_{AB}.

12.10   In the boundary layer over a solid surface, the fluid velocity and temperature profiles can be approximated by

$$u(y) = Ay + By^2 - Cy^3 \qquad T(y) = D + Ey + Fy^2 - Gy^3$$

where y is the axis orthogonal to the surface and the rest of the coefficients are constants. Obtain an expression for the friction coefficient f and the heat transport coefficient h as a function of the above constants, U_∞, T_∞ and the fluid properties.

12.11   A way to keep a liquid cool at high ambient temperatures consists of covering its container with a damp cloth, like felt. This principle is applied, for example, to water bottles. Assume that the container is exposed

at an atmosphere of dry air at 40 °C. The cloth surrounding the container is moistened with a liquid of 200 kg/kmol molar mass and 100 kJ/kg latent heat of vaporization. The saturation pressure at those conditions is $P_s = 5\,000$ Pa and the diffusion coefficient of the vapor in air, $D = 0.2 \times 10^{-4}$ m^2/s. What is the container temperature and that of the liquid that it contains?
Data for air: $\mu = 1.8 \times 10^{-5}$ kg/(m s), $c_p = 1\,007$ J/(kg K), $\kappa = 0.026$ W/(m K), $\rho = 1.2$ kg/m^3.

12.12   On a cold day in April a jogger losses $2\,000$ W due to convective heat transfer between the jogger's skin, which is maintained dry at a temperature of 30 °C, and the environment, also dry, at a temperature of 10 °C. Three months later, the jogger moves at the same pace but the day is warm and humid, with a temperature of $T_\infty = 30$ °C and a relative humidity of $\Phi = 60\%$. The skin of the jogger is sweating and at a temperature of 35 °C. In both cases, the properties of air can be considered constant and equal to: $\nu = 1.6 \times 10^{-5}$ m^2/s, $\kappa = 0.026$ W/(m K), Pr $= 0.7$, $D = 2.3 \times 10^{-5}$ m^2/s (for water vapor in air), $L = 2257$ kJ/kg (latent heat of vaporization), $P_s = 6221$ Pa.

(a) What is the rate of water evaporation on a summer day?
(b) What is the total heat loss per unit time during the summer day?

12.13   Cooling and heating involved in boiling and condensation processes depend on the fluid properties (ρ, μ, κ, c_p), a characteristic length L, a characteristic temperature difference ΔT, on the characteristic buoyance force between the liquid and gas phases ($\rho_{\text{liq}} - \rho_{\text{vap}})g$, the latent heat of vaporization h_{lv} and the surface tension σ. Determine the dimensionless parameters that govern the behavior of the dimensionless heat transport coefficient Nu.

12.14   Check the dimensionless expression (12.60) for the natural convection mass transport coefficient.

12.15   Check the derivation of expression (12.54).

Part V

Self Evaluation

13

Self Evaluation Exercises

Problems

13.1 An axle of diameter d turns inside a fixed concentric bearing of diameter D. The length of the device is L. The space between the axle and the bearing is filled with an oil of viscosity μ. The axle turns at an angular velocity ω so that in the steady state the fluid velocity has only a tangential direction e_θ and it is a quadratic function of the radius, with a minimum where the velocity is zero.

(a) How much is the heat per unit time $\dot{Q}$ to be eliminated from the device so the fluid is maintained at constant temperature?

(b) If the device is isolated so $\dot{Q} = 0$, assuming the equation of state $de = c_v dT$, what is the rate of variation of the temperature ?

13.2 Given the two-dimensional velocity field

$$v = \left\{ \begin{array}{c} 5y^2 \\ 3y - 3 \end{array} \right\}$$

(a) Calculate the divergence $\nabla \cdot v$.

(b) Is the flow compressible or incompressible? Why?

(c) Determine the viscous dissipation function ϕ_v.

13.3 An approximate method to scale cylindrical stirring tanks for liquids consists of maintaining the power per unit volume $p_v = P/V$ constant. It is considered that the agitation power P is a function of the diameter of the agitator D, its angular velocity ω and the liquid density ρ.

(a) Determine the dimensionless relation of P with respect to the other dimensionless variables.

(b) It is desired to increase the tank volume by 3. What is the scale factor of the diameter and the agitator?

(c) What is the power and the angular velocity for the new tank?

Note. Assume that the tanks are geometrically similar and the flow is turbulent.

13.4   Determine the horizontal and vertical forces to fix the elbow of the Figure.

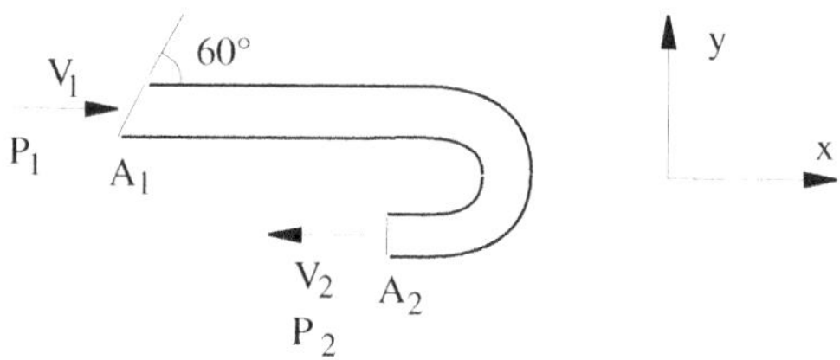

Problem 13.4. Force to hold a 180° elbow.

13.5   The Figure shows the velocity v_x and temperature T profiles around a cylinder. The ambient pressure far from the cylinder can be taken constant and equal to 0. If the flow is steady and incompressible, determine the following variables.

(a) The mass flow $\dot{m}$ across the horizontal surfaces of the control volume.
(b) The force F_D necessary to keep the cylinder (of length L) fixed. Calculate the dimensionless friction coefficient, $C_f = \bar{\tau}_0/(\frac{1}{2}\rho U_0^2)$, where the mean stress is defined as $\bar{\tau}_0 = F_D/(2\pi DL)$.
(c) If $de = c_v dT$ and the cylinder is heated at a rate of $\dot{Q}$ calculate T_s for the temperature profiles shown in the Figure. Assume that c_v is constant.

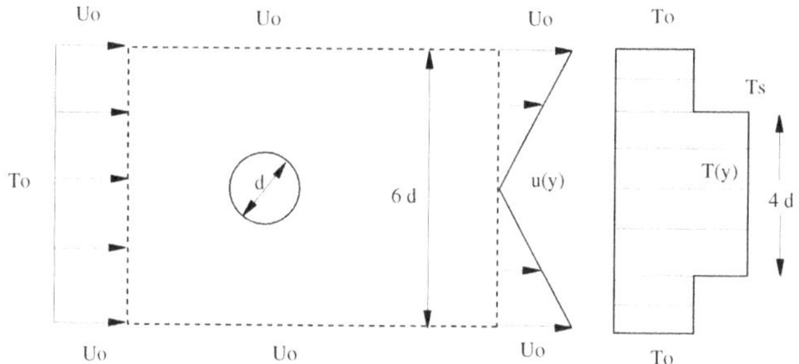

Problem 13.5. Non-isothermal incompressible flow about a cylinder.

13.6   Calculate the vertical F_1 and horizontal F_2 net forces and the point of application y_2 to hold the wall of the tank.

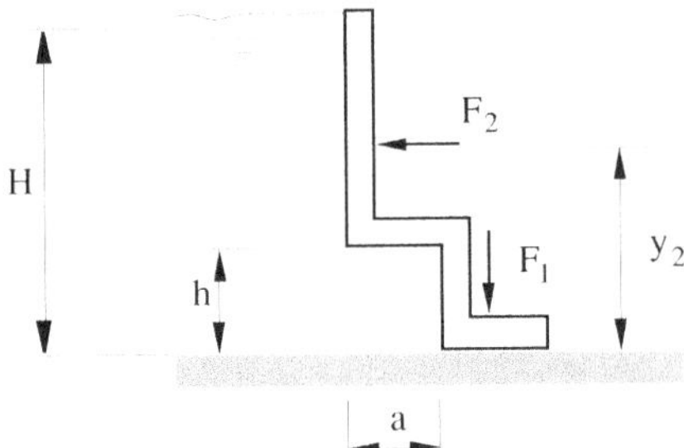

Problem 13.6. Force to hold a tank wall.

13.7 Let a non-Newtonian fluid have the constitutive equation

$$\tau' = \mu \left| \frac{du}{dy} \right|^{n} \qquad n > 1$$

(a) To what kind of non-Newtonian fluid does it correspond?
(b) If the velocity profile near a solid wall can be expressed as

$$u(y) = a_1 y + a_2 y^2$$

where y is the coordinate orthogonal to the wall, determine τ'_0, the viscous stress at the wall.

13.8 A circular container of 230 mm diameter filled with water at ambient temperature loses mass at a rate of 1.5×10^{-5} kg/s when the ambient is dry and at 22 °C.

(a) Determine the mass transfer coefficient
(b) Calculate the total heat (by convection and evaporation) which is lost when the ambient air has a relative humidity of 50% and the water temperature is 37 °C.
    Gas constant: $R = 8.314$ kJ/(kmol K). *Water properties:* $D_{AB} = 2.3 \times 10^{-5}$ m^2/s; *latent heat of vaporization* $L = 2\,257$ kJ/kg. *Air properties:* $\rho = 1.2$ kg/m^3; $\mu = 1.82 \times 10^{-5}$ kg/(m s); $\kappa = 0.026$ W/(m K); Pr $= 0.7$.

13.9 The Figure of the problem shows a common technique to disperse fluid B into fluid A to form the solution AB. The technique consists of mixing both substances through concentric pipes.

(a) Calculate u_2^{AB} and the pressure loss assuming negligible friction forces and equal densities for all the fluids. Characterize the result for $D = 10$ cm, $d = 2$ cm, $\rho_A = \rho_B = \rho_{AB} = 1$ gr/cm^3, $u_1^A = 1.5$ m/s, $u_1^B = 4.0$ m/s.
(b) Obtain the pressure loss for the case of different densities as a function of d/D, ρ_A/ρ_B, u_1^A/u_1^B and $\rho_A u_1^A/\rho_B u_1^B$.

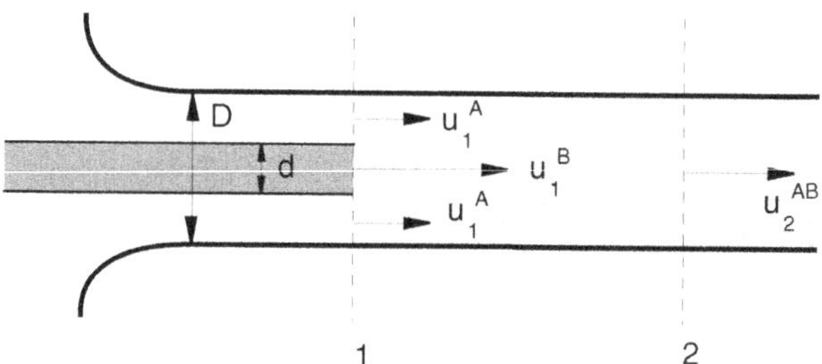

Problem 13.9. Mixing process through concentric pipes.

(c) What can be concluded from the expression obtained in (b)?

13.10   The conic pivot of the Figure spins at an angular velocity ω and rests over a thin layer of oil with thickness h. Determine the moment due to viscous friction as a function of the angle α, the viscosity μ, the angular velocity, the thickness h and the diameter of the axle D.

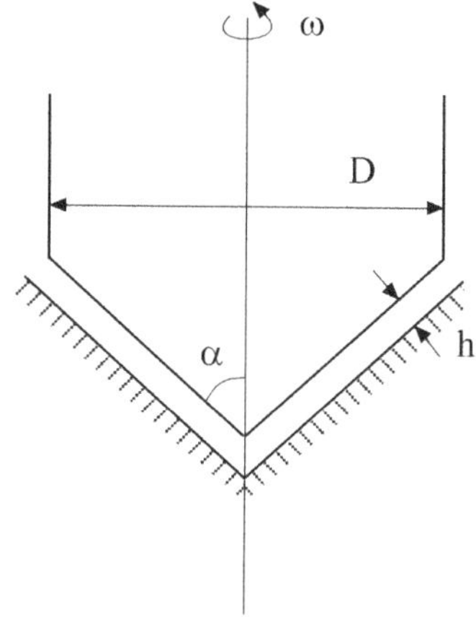

Problem 13.10. Conic bearing to support axial and radial forces.

13.11   From a vertical tube of length L and radius R, a fluid of density ρ and viscosity μ falls. Assuming that the velocity profile is steady, fully developed and parabolic,

$$v(r) = \frac{2Q}{\pi R^4} \left(R^2 - r^2 \right)$$

and that the gravity acts downwards, determine the outgoing volumetric flux Q.

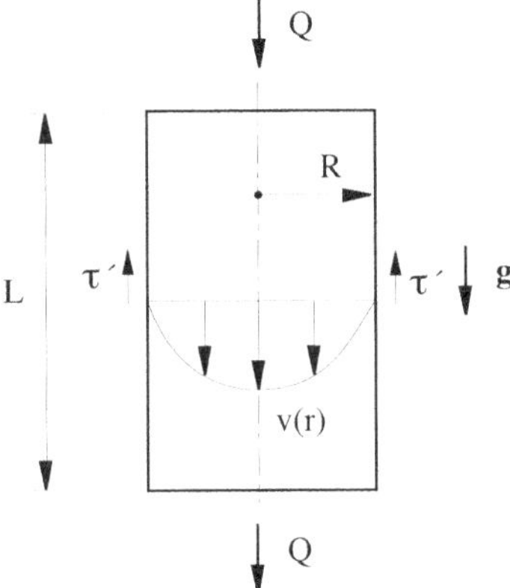

Problem 13.11. Fall of a fluid in a pipe due to gravity.

13.12 The evolution of small perturbations in a fluid can be modeled by the Orr-Sommerfeld equation

$$\nu\frac{\mathrm{d}^4\phi(y)}{\mathrm{d}y^4} - \left[\omega + 2\nu k^2\right]\frac{\mathrm{d}^2\phi(y)}{\mathrm{d}y^2} + k^2\left[\omega + \nu k^2\right]\phi(y) = 0$$

where ϕ is the stream function $[\mathrm{m}^2/\mathrm{s}]$, ω the angular velocity of the wave $[1/\mathrm{s}]$, k the wavenumber $[1/\mathrm{m}]$ and $\nu = \mu/\rho$, the kinematic viscosity. Using ρ, μ, h, U, make the Orr-Sommerfeld equation dimensionless.

13.13 The time t to discharge a tank depends approximately on its diameter D, the liquid level h, the diameter of the outlet orifice d, the acceleration of the gravity g and the fluid density ρ.

(a) Determine the dimensionless relation for the discharge time.
(b) If a tank is made at a scale four times smaller, what is the discharge time? How much should h be for dimensional analysis to apply?
(c) If the fluid is changed, what is the time of discharge? Justify the answer.
(d) In this section, the viscous effects μ are taken into account. What is the new dimensionless number that appears in the nondimensional relation? Is it possible to have complete similarity when both, the geometric scale of the tank and the fluid are modified? Why?

13.14 A porous cylinder of unknown surface is saturated with water. Dry air is blown perpendicularly to the cylinder at a pressure of 1 atm and velocity 10 m/s, so the air gets humid. The water evaporation rate is 1.684×10^{-5} kg/s and the heat transport coefficient is given by

$$\mathrm{Nu}_D = C\ \mathrm{Re}_D^m\ \mathrm{Pr}^{1/3} \qquad \text{where } C = 0.193 \text{ and } m = 0.618$$

Calculate the surface of the cylinder assuming that both the water and air are at 310 K and that the cylinder diameter is 0.045 m.

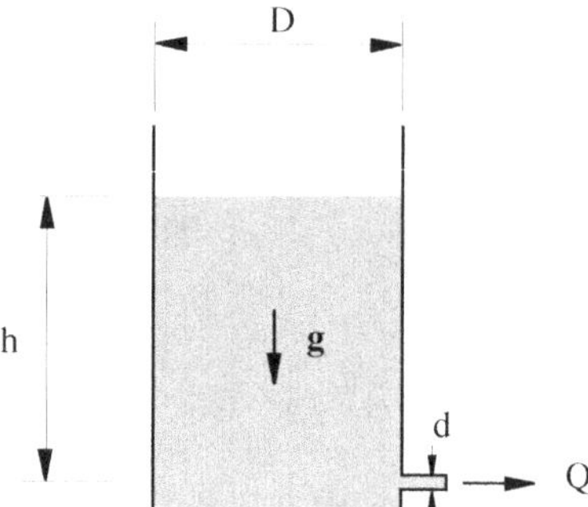

Problem 13.13. Draining of a tank through an orifice. Dimensional analysis.

Data for air: $\rho = 1.2$ kg/m^3; $\mu = 1.82 \times 10^{-5}$ kg/(m s); $D_a = 2.3 \times 10^{-5}$ m^2/s; Pr $= 0.7$. *Gas constant:* $R = 8.314$ kJ/(kmol K). *Water vapor properties:* $M_a = 18$ kg/kmol

13.15   Given the two-dimensional velocity field

$$v = \left\{ \begin{array}{c} -2x^2t \\ 6yt \end{array} \right\}$$

determine the equation of the streakline that passes by the point (x_0, y_0).

13.16   A planet in formation is made of a fluid of constant density ρ. If at the free surface, located at $r = R$, there is atmospheric pressure p_{atm} and the radial body forces are $f_m = -\frac{4\pi K}{3}r$, determine the hydrostatic pressure distribution $p(r)$ inside the planet.

13.17   In a wind tunnel there is a uniform air flow of 7 m/s (kinematic viscosity $\nu = \mu/\rho = 1.5 \times 10^{-5}$ m^2/s) at 295 K. Aligned with the flow, there is a 4 m long rectangular container, filled with water to a height of 1 cm. If the vapor pressure at the ambient conditions is 2000 Pa and the water is at the air temperature, calculate the time to evaporate the water in the container. The global mass transfer coefficient can be approximated by

$$\overline{\text{Sh}_L} = 0.664\,\text{Re}_L^{1/2}\text{Sc}^{1/3}$$

Gas constant: $R = 8\,314$. J/(kmol K). *Mass diffusivity of water in air:* $D_A = 2.5 \times 10^{-5}$m^2/s.

13.18   The impulsion power of a hydraulic pump is frequently expressed as a function of energy head H [m]. It can be shown that gH (with g the gravity acceleration) depends on the fluid density, ρ [kg/m^3], and viscosity μ [kg/(m s)], the pump angular velocity of rotation N [rad/s], the runner diameter D [m], the volumetric flow rate Q [m^3/s] and the characteristic roughness length ϵ [m].

(a) Determine the characteristic curve gH in dimensionless form, using as fundamental variables ρ, N and D.
(b) Assume that the dissipation effects are negligible, that is, ignore the variables μ and ϵ. In this new situation, called partial similarity, what is the dimensionless relation?
(c) Assume a pump with characteristic curve $H(Q) = 20 - 0.1Q^2$. Assuming the partial similarity of (b), what would the new characteristic curve of the pump $H'(Q')$ be if the rotation speed was doubled?
(d) Again, neglecting the dissipation effects, what is the new characteristic curve if only the fluid density is modified?

13.19 The Figure shows a two-dimensional adiabatic mixing tank. If the flow is steady and incompressible (with density ρ), answer the following questions.

(a) Calculate the exit volumetric flow rate Q_3.
(b) Given p_1, p_2 and p_3, employ the mechanical energy equation to determine the viscous dissipation in the tank D_v.
(c) As a function of the inlet temperatures, T_1, T_2, and the specific heat at constant volume c_v (which can be assumed constant), calculate the exit temperature T_3.

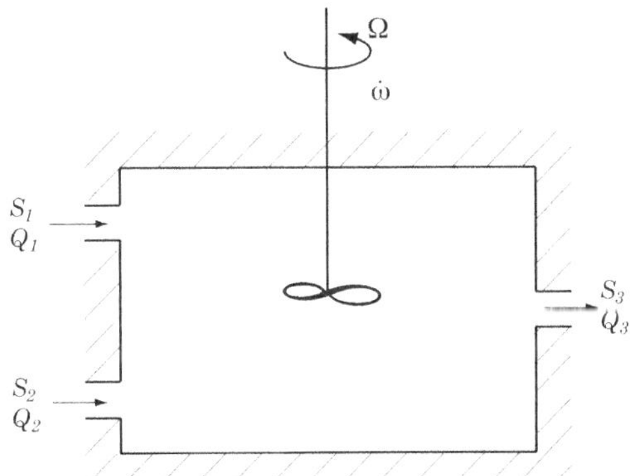

Problem 13.19. Mixing tank.

13.20 A vertical solar panel is $L = 1$ m tall and $w = 2$ m wide. The local Nusselt number follows the correlation

$$\mathrm{Nu}_x = C\left(\frac{\mathrm{Gr}_x}{4}\right)^{1/4} \qquad C = 0.56$$

where the Grashof number is defined as

$$\mathrm{Gr}_L = \frac{\beta g \rho^2 L^3 \Delta T}{\mu^2}$$

with β the expansion coefficient and ΔT the temperature difference between the panel and the environment. Determine the correlation for the global Nusselt number Nu_L.

Part VI

Appendices

A

Collection of Formulae

A.1 Integral Equations for a Control Volume

A.1.1 Mass Conservation Equation

$$\frac{\mathrm{d}}{\mathrm{d}t} \int_{V_c(t)} \rho \, \mathrm{d}V + \int_{S_c(t)} \rho \left[(\boldsymbol{v} - \boldsymbol{v}^c) \cdot \boldsymbol{n} \right] \mathrm{d}S = 0 \tag{A.1}$$

A.1.2 Chemical Species Conservation Equation

$$\frac{\mathrm{d}}{\mathrm{d}t} \int_{V_c(t)} \rho_A \, \mathrm{d}V + \int_{S_c(t)} \rho_A \left[(\boldsymbol{v} - \boldsymbol{v}^c) \cdot \boldsymbol{n} \right] \mathrm{d}S$$
$$= - \int_{S_c(t)} \rho_A (\boldsymbol{v}_A - \boldsymbol{v}) \cdot \boldsymbol{n} \, \mathrm{d}S + \int_{V_c(t)} \dot{\omega}_A \, \mathrm{d}V \tag{A.2}$$

A.1.3 Momentum Equation

$$\frac{\mathrm{d}}{\mathrm{d}t} \int_{V_c(t)} \rho \boldsymbol{v} \, \mathrm{d}V + \int_{S_c(t)} \rho \boldsymbol{v} \left[(\boldsymbol{v} - \boldsymbol{v}^c) \cdot \boldsymbol{n} \right] \mathrm{d}S$$
$$= \int_{S_c(t)} (-p\boldsymbol{n} + \boldsymbol{\tau}' \cdot \boldsymbol{n}) \, \mathrm{d}S + \int_{V_c(t)} \rho \boldsymbol{f}_m \, \mathrm{d}V \tag{A.3}$$

A.1.4 Angular Momentum Equation

$$\frac{\mathrm{d}}{\mathrm{d}t} \int_{V_c(t)} \boldsymbol{r} \times \rho \boldsymbol{v} \, \mathrm{d}V + \int_{S_c(t)} \boldsymbol{r} \times \rho \boldsymbol{v} \left[(\boldsymbol{v} - \boldsymbol{v}^c) \cdot \boldsymbol{n} \right] \mathrm{d}S$$
$$= \int_{S_c(t)} \boldsymbol{r} \times (\boldsymbol{\tau} \cdot \boldsymbol{n}) \, \mathrm{d}S + \int_{V_c(t)} \boldsymbol{r} \times \rho \boldsymbol{f}_m \, \mathrm{d}V \tag{A.4}$$

A.1.5 Mechanical Energy Equation

As a Function of Body Forces

$$\frac{\mathrm{d}}{\mathrm{d}t} \int_{V_c(t)} \rho \frac{1}{2} v^2 \, \mathrm{d}V + \int_{S_c(t)} \rho \frac{1}{2} v^2 \left[(\boldsymbol{v} - \boldsymbol{v}^c) \cdot \boldsymbol{n} \right] \mathrm{d}S$$

$$= \int_{S_c(t)} (-p\boldsymbol{n} + \boldsymbol{\tau}' \cdot \boldsymbol{n}) \cdot \boldsymbol{v} \, \mathrm{d}S + \int_{V_c(t)} p \boldsymbol{\nabla} \cdot \boldsymbol{v} \, \mathrm{d}V \qquad (A.5)$$

$$- \int_{V_c(t)} \phi_v \, \mathrm{d}V + \int_{V_c(t)} \rho \boldsymbol{f}_m \cdot \boldsymbol{v} \, \mathrm{d}V$$

As a Function of Potential Energy

$$\frac{\mathrm{d}}{\mathrm{d}t} \int_{V_c(t)} \rho \left(\frac{1}{2} v^2 + U \right) \mathrm{d}V + \int_{S_c(t)} \rho \left(\frac{1}{2} v^2 + U \right) \left[(\boldsymbol{v} - \boldsymbol{v}^c) \cdot \boldsymbol{n} \right] \mathrm{d}S$$

$$= \int_{S_c(t)} (-p\boldsymbol{n} + \boldsymbol{\tau}' \cdot \boldsymbol{n}) \cdot \boldsymbol{v} \, \mathrm{d}S + \int_{V_c(t)} p \boldsymbol{\nabla} \cdot \boldsymbol{v} \, \mathrm{d}V - \int_{V_c(t)} \phi_v \, \mathrm{d}V$$

$$U = gz$$

$$(A.6)$$

A.1.6 Total Energy Equation

$$\frac{\mathrm{d}}{\mathrm{d}t} \int_{V_c(t)} \rho \left(e + \frac{1}{2} v^2 + U \right) \mathrm{d}V$$

$$+ \int_{S_c(t)} \rho \left(e + \frac{1}{2} v^2 + U \right) \left[(\boldsymbol{v} - \boldsymbol{v}^c) \cdot \boldsymbol{n} \right] \mathrm{d}S$$

$$= \int_{S_c(t)} (-p\boldsymbol{n} + \boldsymbol{\tau}' \cdot \boldsymbol{n}) \cdot \boldsymbol{v} \, \mathrm{d}S - \int_{S_c(t)} \boldsymbol{q} \cdot \boldsymbol{n} \, \mathrm{d}S + \int_{V_c(t)} \dot{q}_v \, \mathrm{d}V$$

$$(A.7)$$

A.1.7 Internal Energy Equation

$$\frac{\mathrm{d}}{\mathrm{d}t} \int_{V_c(t)} \rho e \, \mathrm{d}V + \int_{S_c(t)} \rho e \left[(\boldsymbol{v} - \boldsymbol{v}^c) \cdot \boldsymbol{n} \right] \mathrm{d}S$$

$$= - \int_{S_c(t)} p \boldsymbol{\nabla} \cdot \boldsymbol{v} \, \mathrm{d}V + \int_{V_c(t)} \phi_v \, \mathrm{d}V \qquad (A.8)$$

$$- \int_{S_c(t)} \boldsymbol{q} \cdot \boldsymbol{n} \, \mathrm{d}S + \int_{V_c(t)} \dot{q}_v \, \mathrm{d}V$$

A.2 Relevant Dimensionless Numbers

Table A.1. Relevant dimensionless numbers in transport phenomena.

Number	Definition	Physics	Equivalence
Strouhal	$S = \frac{L}{vt}$	$\dfrac{\text{residence time}}{\text{characteristic time}}$	
Reynolds	$Re = \frac{\rho v L}{\mu}$	$\dfrac{\text{viscous convection}}{\text{viscous diffusion}}$	$= \frac{vL}{\nu}$
Péclet	$Pe = \frac{\rho c_p v L}{\kappa}$	$\dfrac{\text{thermal convection}}{\text{thermal diffusion}}$	$= \frac{vL}{\alpha}$
Péclet II	$Pe_{II} = \frac{vL}{D_{AB}}$	$\dfrac{\text{mass convection}}{\text{mass diffusion}}$	$= \frac{vL}{D_{AB}}$
Damköhler I	$Da_{I} = \frac{\dot{\omega}_A L}{\rho_A v}$	$\dfrac{\text{chemical generation}}{\text{mass convection}}$	$= \frac{vL}{D_{AB}}$
Prandtl	$Pr = \frac{\mu c_p}{\kappa}$	$\dfrac{\text{dynamic diffusion}}{\text{thermal diffusion}}$	$= \frac{Pe}{Re} = \frac{\nu}{\alpha}$
Schmidt	$Sc = \frac{\mu}{\rho D_{AB}}$	$\dfrac{\text{dynamic diffusion}}{\text{mass diffusion}}$	$= \frac{Pe_{II}}{Re} = \frac{\nu}{D_{AB}}$
Lewis	$Le = \frac{\kappa}{\rho c_p D_{AB}}$	$\dfrac{\text{thermal diffusion}}{\text{mass diffusion}}$	$= \frac{Sc}{Pr} = \frac{\alpha}{D_{AB}}$
Drag coefficient	$C_D = \frac{\tau_0'}{\frac{1}{2}\rho v^2}$	$\dfrac{\text{friction}}{\text{dynamic pressure}}$	$= C_f$
Nusselt	$Nu = \frac{hl}{\kappa}$	heat transport coeff.	
Stanton	$St = \frac{h}{\rho c_p v}$	heat transport coeff.	$= \frac{Nu}{Pe}$
Sherwood[1]	$Sh = \frac{h_m l}{D_{AB}}$	mass transport coeff.	
Mass Stanton	$St_m = \frac{h_m}{v}$	mass transport coeff.	$= \frac{Sh}{Pe_{II}}$
Grashof	$Gr = \frac{\rho^2 g \beta \Delta T L^3}{\mu^2}$	$\dfrac{\text{buoyancy forces}}{\text{viscous forces}}$	
Rayleigh	$Ra = \frac{\rho g \beta \Delta T L^3}{\mu \alpha}$	$\dfrac{\text{buoyancy forces}}{\text{thermal diffusion}}$	$= Gr\,Pr$
Weber	$We = \frac{\rho v^2 L}{\sigma}$	$\dfrac{\text{inertial forces}}{\text{line forces}}$	

[1] Also Nusselt for mass transfer, Nu_m

Table A.3. Dimensionless numbers encountered in the transport equations.

	Temp./Conv.	Conv./Diff.	Gen./Conv.	Conv./Grav.
Mass cons.	$S = \frac{l/v}{t}$			
Momentum	$S = \frac{l/v}{t}$	$Re = \frac{\rho v l}{\mu}$	$Eu = \frac{\Delta p}{\rho v^2}$	$Fr = \frac{v^2}{gl}$
Energy	$S = \frac{l/v}{t}$	$Pe = \frac{\rho c_p v l}{\kappa}$		
Chem. esp. A	$S = \frac{l/v}{t}$	$Pe_{II} = \frac{vl}{D_A}$	$Da_I = \frac{l \dot\omega_A}{\rho_A v}$	

A.3 Transport Coefficient Analogies

A.3.1 Analogy of Reynolds

$$\frac{Re}{2} \, C_f = Nu = Sh \qquad Pr = 1 \tag{A.9}$$

A.3.2 Analogy of Chilton-Colburn

$$\begin{aligned}
\tfrac{1}{2}\, C_f &= St\, Pr^{2/3} & 0.6 < Pr < 60 \\
\tfrac{1}{2}\, C_f &= St_m\, Sc^{2/3} & 0.6 < Sc < 6000
\end{aligned} \tag{A.10}$$

$$Sh = Nu\, Le^{1/3} \qquad \left\{ \begin{array}{l} 0.6 < Pr < 60 \\ 0.6 < Sc < 6000 \end{array} \right. \tag{A.11}$$

B

Classification of Fluid Flow

Fluid flow can be classified according to the following criteria.

B.1 Stationary (steady) / non-stationary (transient, periodic)

The flow is *stationary* when the fluid variables in the Eulerian description do not depend on time. That is, for any Eulerian variable the partial derivative with respect to time is zero,

$$\frac{\partial \cdot}{\partial t} = 0 \tag{B.1}$$

Example B.1. The axial velocity in a circular section, straight pipe of radius R, when it becomes fully developed, for laminar flow, can be expressed as

$$v(r) = V_0 \left[1 - \left(\frac{r}{R} \right)^2 \right]$$

Because $v(r)$ does not depend on time t, it is a steady or stationary flow. This flow is called Hagen-Poiseuille flow.

When this is not the case, the flow is *transient* or *unsteady*. This is the situation when the flow is evolving from the initial state towards the stationary solution.

Sometimes the long term solution is not steady, but it repeats itself in time. This flow is called *periodic*, as the flow around a circular cylinder at $Re = 100$.

B.2 Compressible / incompressible

All substances are *compressible* to a certain extent, i.e. of variable density. However, in many practical situations, the density variations are so small that they can be neglected and the density can be considered constant.

Flows that are modeled assuming constant density are called *incompressible*, and is typical of liquids. An incompressible flow satisfies

$$\nabla \cdot \boldsymbol{v} = \operatorname{div} \boldsymbol{v} = 0 \tag{B.2}$$

Compressible flows can be classified as subsonic, transonic or supersonic, depending on the Mach number Ma, ratio between the local fluid velocity and the local propagation speed of the sound.

The flow in gases can be considered incompressible if $\mathrm{Ma} < 0.3$.

Example B.2. At $20°$C the speed of sound in air is about 340 m/s. The flow around a vehicle that moves at $100 \text{ km/h} \approx 28 \text{ m/s}$ has a $\mathrm{Ma} = 28/340 \approx 0.08$ and the compressibility effects can be ignored.

B.3 One-dimensional / Two-dimensional / Three-dimensional

The flow is one-, two- or three-dimensional when the fluid variables (such as density, velocity, temperature, etc.) depend on one, two or three spatial coordinates.

There are only a few pure one-dimensional flows. Most of them are two- and three-dimensional.

Plane flow is a two-dimensional flow with respect to Cartesian or polar coordinates. An *axisymmetric* flow is a two-dimensional flow in cylindrical coordinates. A spherically symmetric flow is an example of one-dimensional flow.

Example B.3 (One-dimensional flow). The incompressible flow out of a sink is a composition of a radial plus a tangential velocity component and can be modeled in cylindrical coordinates as

$$\begin{aligned}
v_r(r) &= \frac{Q}{2\pi r} \\
v_\theta(r) &= \frac{Q}{2\pi r \tan \alpha} \\
v_z(r) &= 0
\end{aligned}$$

where Q and α are constants. This is a one-dimensional incompressible flow.

Example B.4 (Two-dimensional flow). In the flow around an infinitely long cylinder placed orthogonally to the stream, the flow field is the same in all the planes normal to the cylinder. Thus, this is a plane two-dimensional flow field.

B.4 Viscous / Ideal

In real fluids, due to the friction between the layers of fluid, there is a resistance to the motion. This friction is caused by a fluid property called *viscosity μ*. Flows where friction is taken into account are termed *viscous flows*.

In *ideal* flows this property is neglected, $\mu = 0$, giving rise to frictionless fluid motions. For this type of fluid, the rest of the diffusion coefficients are neglected, like thermal conductivity $\kappa = 0$ and mass diffusivity $D_{AB} = 0$.

B.5 Isothermal / Adiabatic

In an isothermal flow, the temperature is constant. The opposite is a non-isothermal flow.

Adiabatic means that the system is thermally isolated and there is no heat transport between itself and its surroundings.

B.6 Rotational / Irrotational

A flow is said to be *irrotational* if the fluid particles along their path translate without rotation about the particle center. For an irrotational flow

$$\text{curl}\, \boldsymbol{v} = \boldsymbol{0} \tag{B.3}$$

If along their path the fluid particles translate and rotate about the particle center, the flow is *rotational*.

B.7 Laminar / Turbulent

The flow is *laminar* when the motion of the fluid particles is well-organized, as if layers of fluid slide over others. It is predictable and deterministic.

However, due to the nonlinearities of the transport equations, a fluid flow can have a random component, so that the real flow is the sum of an average motion plus some chaotic fluctuations. This flow is called *turbulent* and it is always three-dimensional and unsteady. Another relevant trait of turbulent flow compared to laminar flow is that it enhances the transport of momentum, heat and mass.

Most of the industrial and natural flows are turbulent.

C

Substance Properties

C.1 Properties of water

Table C.1. Properties of pure water at atmospheric pressure [15].

T	ρ	β	μ	ν	α	c_p	Pr
°C	kg/m^3	K^{-1}	kg/(m s)	m^2/s	m^2/s	J/(kg K)	ν/α
0	1000	-0.6E-4	1.788E-3	1.788E-6	1.33E-7	4217	13.4
10	1000	0.9E-4	1.307E-3	1.307E-6	1.38E-7	4192	9.5
20	998	2.1E-4	1.003E-3	1.005E-6	1.42E-7	4182	7.1
30	996	3.0E-4	0.799E-3	0.802E-6	1.46E-7	4178	5.5
40	996	3.8E-4	0.657E-3	0.662E-6	1.52E-7	4178	4.3
50	998	4.5E-4	0.548E-3	0.555E-6	1.58E-7	4180	3.5

Latent heat of evaporation at 100 °C $= 2.257 \times 10^6$ J/kg
Latent heat of fusion of ice at 0 °C $= 0.334 \times 10^6$ J/kg
Ice density 920 kg/m^3
Surface tension between water and air at 20 °C $= 0.0728$ N/m

C.2 Properties of dry air at atmospheric pressure

At 20 °C and 1 atm, $c_p = 1\,012$ J/(kg K)
At 20 °C and 1 atm, $c_v = 718$ J/(kg K)

Table C.2. Saturation curve of water vapor in air.

T_{sat}	p_{sat}
K	Pa
295	2 617
300	3 531
310	6 221
320	10 530
325	13 510

Table C.3. Properties of dry air at atmospheric pressure.

T	ρ	μ	ν	α	Pr
°C	kg/m^3	kg/(m s)	m^2/s	m^2/s	ν/α
0	1.293	1.71E-5	1.33E-5	1.84E-5	0.72
10	1.247	1.76E-5	1.41E-5	1.96E-5	0.72
20	1.200	1.81E-5	1.50E-5	2.08E-5	0.72
30	1.165	1.86E-5	1.60E-5	2.25E-5	0.71
40	1.127	1.90E-5	1.69E-5	2.38E-5	0.71
60	1.060	2.00E-5	1.88E-5	2.65E-5	0.71
80	1.000	2.09E-5	2.09E-5	2.99E-5	0.70
100	0.946	2.18E-5	2.30E-5	3.28E-5	0.70

D

A Brief Introduction to Vectors, Tensors and Differential Operators

D.1 Indicial Notation

Physical quantities are represented by tensors. Tensors are members of a set of entities that encompasses scalars, vectors and more complicated variables. The number of indices in a tensor is called the *order* or *rank* of the tensor. Examples of tensors are

(a) A scalar T, a tensor of order 0.
(b) A vector v, a tensor of order 1.
(c) Typically, the word tensor is used to denote a second-order tensor, a tensor with two indices. Second-order tensors are matrices that transform in a physical way under changes of coordinate systems.

Since tensors have indices, instead of using *tensor notation*, it is very convenient to use *indicial notation*, that is, exposing their indices (subscripts and/or superscripts). This way, the equations take on a more compact form.

Example D.1 (Tensors). In indicial notation a vector v

$$v = \left\{ \begin{array}{c} v_1 \\ v_2 \\ v_3 \end{array} \right\} \tag{D.1}$$

is denoted as v_i, $i = 1, 2, 3$. A second-order tensor m,

$$m = \begin{bmatrix} m_{11} & m_{12} & m_{13} \\ m_{21} & m_{22} & m_{23} \\ m_{31} & m_{32} & m_{33} \end{bmatrix} \tag{D.2}$$

is denoted as m_{ij}, $i, j = 1, 2, 3$.

Usually, together with indicial notation, the *Einstein summation convention* on repeated indices is employed: In a term, when a index is repeated, then there is an implied sum on that index.

Example D.2 (Einstein summation convention). The scalar product of two vectors in Cartesian coordinates is

$$a_i b_i = \sum_{i=1}^{3} a_i b_i = a_1 b_1 + a_2 b_2 + a_3 b_3$$

On the left-hand side, in the term $a_i b_i$ the index i appears twice. Therefore, there is an implied summation on i, $i = 1, 2, 3$.

Example D.3 (Einstein summation convention). The i-th component of the matrix-vector product $\boldsymbol{ma}$ is

$$(\boldsymbol{ma})_i = \sum_{j=1}^{3} m_{ij} a_j = m_{ij} a_j \qquad (i = 1, 2, 3)$$

On the right-hand side, the sum symbol has been eliminated because in $m_{ij} a_j$ the index j is repeated, indicating addition on j. The index i is free, indicating that the product is a first-order tensor, that is, a vector.

Two important tensors are the Kronecker delta, δ_{ij}, and the cyclic or permutation tensor ϵ_{ijk}.

Definition D.1 (Kronecker delta). *The tensor Kronecker delta, δ_{ij}, is the unit tensor, which takes the unit value for $i = j$ and zero for $i \neq j$. Thus,*

$$\boldsymbol{\delta} = \begin{bmatrix} 1 & 0 & 0 \\ 0 & 1 & 0 \\ 0 & 0 & 1 \end{bmatrix} \tag{D.3}$$

Example D.4 (Kronecker delta). Calculate $v_i \delta_{ij} v_j$ where

$$\boldsymbol{v} = \left\{ \begin{array}{c} 1 \\ 2 \\ 3 \end{array} \right\}$$

Solution. The term to be computed has two indices, i, j, which are repeated and so, imply summation. Therefore,

$$
\begin{aligned}
v_i \delta_{ij} v_j \; &= \textstyle\sum_{i,j=1}^{3} v_i \delta_{ij} v_j \\
&= v_1 \delta_{11} v_1 + v_1 \delta_{12} v_2 + v_1 \delta_{13} v_3 \\
&\quad + v_2 \delta_{21} v_1 + v_2 \delta_{22} v_2 + v_2 \delta_{23} v_3 \\
&\quad + v_3 \delta_{31} v_1 + v_3 \delta_{32} v_2 + v_3 \delta_{33} v_3 \\
&= v_1 \delta_{11} v_1 + v_2 \delta_{22} v_2 + v_3 \delta_{33} v_3
\end{aligned}
$$

Taking into account that δ_{ij} is non-zero only if $i = j$, the expression can be written directly as

$$
\sum_{i,j=1}^{3} v_i \delta_{ij} v_j = \sum_{i} v_i \delta_{ii} v_i
$$

Furthermore, the diagonal components of the Kronecker delta are the unit, so

$$
\begin{aligned}
v_i \delta_{ij} v_j \; &= v_i \delta_{ii} v_i \\
&= v_i v_i \\
&= v_1^2 + v_2^2 + v_3^2 \\
&= 14
\end{aligned}
$$

Definition D.2 (Permutation tensor). *The permutation tensor ϵ_{ijk} is a third-order tensor which takes on the value 1 when the indices are cyclic ($ijk = 123,\ 231,\ 312$), -1 if the indices are anti-cyclic ($ijk = 321,\ 132,\ 213$), and 0 if there is a repeated index.*

Example D.5 (Permutation tensor). The vector product can be written with the permutation tensor:

$$
(\boldsymbol{a} \times \boldsymbol{b})_i = \epsilon_{ijk} a_j b_k \tag{D.4}
$$

For instance, let us take $i = 1$, the first component of the vector product. Then, the only non-vanishing components are

$$
\begin{aligned}
\epsilon_{1jk} a_j b_k \; &= \epsilon_{123} a_2 b_3 + \epsilon_{132} a_3 b_2 = (+1)a_2 b_3 + (-1)a_3 b_2 \\
&= a_2 b_3 - a_3 b_2
\end{aligned} \tag{D.5}
$$

Likewise,

$$
\epsilon_{2jk} a_j b_k = \epsilon_{231} a_3 b_1 + \epsilon_{213} a_1 b_3 = a_3 b_1 - a_1 b_3 \tag{D.6}
$$

and

$$
\epsilon_{3jk} a_j b_k = \epsilon_{312} a_1 b_2 + \epsilon_{321} a_2 b_1 = a_1 b_2 - a_2 b_1 \tag{D.7}
$$

Compare to Definition D.5.

Definition D.3 (Derivation). *In indicial notation, very frequently derivation with respect to a variable is denoted by the subscript comma followed by the variable or the index that denotes the variable.*

Example D.6 (Derivation). For example, the partial derivative of the temperature T with respect to x can be written as

$$\frac{\partial T}{\partial x} = T_{,x} = T_{,x_1} = T_{,1}$$

D.2 Elementary Vector Algebra

In transport phenomena, we not only have to deal with scalar fields, such as the density or temperature, but also with vectors, like the velocity field. Therefore, let us review some concepts on vector algebra. This section is developed for an orthonormal vector basis.

Consider two vectors $\boldsymbol{a}$ and $\boldsymbol{b}$ with coordinates

$$\boldsymbol{a} = \left\{ \begin{array}{c} a_1 \\ a_2 \\ a_3 \end{array} \right\} \tag{D.8}$$

$$\boldsymbol{b} = \left\{ \begin{array}{c} b_1 \\ b_2 \\ b_3 \end{array} \right\} \tag{D.9}$$

Definition D.4 (Scalar product). *The scalar product of two vectors equals*

$$\begin{aligned} \boldsymbol{a} \cdot \boldsymbol{b} &= a_1 b_1 + a_2 b_2 + a_3 b_3 \\ &= |\boldsymbol{a}||\boldsymbol{b}| \cos \phi \end{aligned} \tag{D.10}$$

with ϕ the angle between the two vectors.

Definition D.5 (Vector product). *The vector product of two vectors is the determinant*

$$\boldsymbol{a} \times \boldsymbol{b} = \det \begin{vmatrix} \boldsymbol{i} & \boldsymbol{j} & \boldsymbol{k} \\ a_1 & a_2 & a_3 \\ b_1 & b_2 & b_3 \end{vmatrix} \tag{D.11}$$

The vector product of two vectors in the plane (x, y) equals

$$\boldsymbol{a} \times \boldsymbol{b} = |\boldsymbol{a}||\boldsymbol{b}| \sin \phi \, \boldsymbol{k} \tag{D.12}$$

Therefore, the vector product of two vectors is another vector, orthogonal to both vectors, with modulus $|\boldsymbol{a}||\boldsymbol{b}| \sin \phi$ where ϕ is the angle between the two vectors.

Example D.7. Let $\boldsymbol{a} = (1,0,0)$ and $\boldsymbol{b} = (1,1,0)$. Determine their scalar and vector product.

Solution. The scalar product is

$$\boldsymbol{a} \cdot \boldsymbol{b} = a_1 b_1 + a_2 b_2 + a_3 b_3$$
$$= 1$$

The vector product could be calculated with a determinant, but since the vectors are in the (x, y) plane

$$\boldsymbol{a} \times \boldsymbol{b} = |\boldsymbol{a}||\boldsymbol{b}| \, \sin \phi \, \boldsymbol{k}$$
$$= 1 \times \sqrt{2} \times \sin 45° \, \boldsymbol{k}$$
$$= 1 \, \boldsymbol{k}$$

Example D.8 (Angle between two vectors). Determine the angle formed by the two vectors of the above example.

Solution. From the scalar product of two vectors, the cosine of the angle between them is

$$\cos \phi = \frac{\boldsymbol{a} \cdot \boldsymbol{b}}{|\boldsymbol{a}||\boldsymbol{b}|} \tag{D.13}$$
$$= \frac{1}{1 \times \sqrt{2}} = \frac{1}{\sqrt{2}}$$

Thus,

$$\phi = 45°$$

Properties.

(a) The scalar product of two perpendicular vectors is zero.
(b) The vector product of two parallel vectors is zero.

D.3 Basic Differential Operators

Continuum mechanics uses differential calculus extensively. In this section, the nabla operator and other basic differential operators are explained. The section is developed for Cartesian coordinates (x, y, z) or (x_1, x_2, x_3).

Definition D.6 (Nabla). *The differential operator nabla* $\boldsymbol{\nabla}$ *is the vector whose components are defined by the spatial partial derivatives*

$$\nabla = \left\{ \begin{array}{c} \dfrac{\partial}{\partial x} \\[1.5ex] \dfrac{\partial}{\partial y} \\[1.5ex] \dfrac{\partial}{\partial z} \end{array} \right\} \tag{D.14}$$

Definition D.7 (Gradient of a scalar field). *The gradient of a scalar field Φ equals $\operatorname{grad}\Phi = \nabla\Phi$. Therefore,*

$$\operatorname{grad}\Phi = \nabla\Phi = \left\{ \begin{array}{c} \dfrac{\partial\Phi}{\partial x} \\[1.5ex] \dfrac{\partial\Phi}{\partial y} \\[1.5ex] \dfrac{\partial\Phi}{\partial z} \end{array} \right\} \tag{D.15}$$

The gradient of a scalar field provides information about the slope of the function in any spatial direction. For instance, the slope of Φ in direction $\boldsymbol{n}$ is

$$\Phi_{,n} = \nabla\Phi \cdot \boldsymbol{n} \tag{D.16}$$

Furthermore the direction of the gradient gives the direction of maximum derivative of Φ and its modulus, the magnitude of this derivate.

Example D.9. Let $h(x,y)$ be the elevation of the ground as a function of the horizontal coordinates (x,y). The gradient ∇h gives the direction of the maximum slope of the ground and its modulus, the slope. The scalar product $\nabla h \cdot \boldsymbol{n}$ reflects the slope in the direction $\boldsymbol{n}$.

Definition D.8 (Divergence of a vector). *The divergence of a vector $\boldsymbol{a}$ equals $\operatorname{div}\boldsymbol{a} = \nabla \cdot \boldsymbol{a}$, that is,*

$$\operatorname{div}\boldsymbol{a} = \nabla \cdot \boldsymbol{a} = \frac{\partial a_1}{\partial x_1} + \frac{\partial a_2}{\partial x_2} + \frac{\partial a_3}{\partial x_3} \tag{D.17}$$

The divergence represents the net local balance of the flux given by the vector $\boldsymbol{a}$ per unit volume of an infinitesimal volume.

Example D.10 (Divergence of the velocity field). For the velocity vector $\boldsymbol{v}$, the divergence is the variation of volume per unit volume and time of a fluid particle. If V denotes the volume of a (infinitesimal) fluid particle,

$$\operatorname{div}\boldsymbol{v} = \frac{1}{V}\frac{dV}{dt} \tag{D.18}$$

Therefore, for an incompressible fluid, $\nabla \cdot \boldsymbol{v} = 0$.

Definition D.9 (Curl of a vector). *The curl of a vector field* $\boldsymbol{a}$ *equals the vector* curl $\boldsymbol{a} = \nabla \times \boldsymbol{a}$*, which is*

$$
\operatorname{curl} \boldsymbol{a} = \nabla \times \boldsymbol{a} = \det \begin{vmatrix} \boldsymbol{i} & \boldsymbol{j} & \boldsymbol{k} \\ \dfrac{\partial}{\partial x_1} & \dfrac{\partial}{\partial x_2} & \dfrac{\partial}{\partial x_3} \\ a_1 & a_2 & a_3 \end{vmatrix}
$$

$$
= \left\{ \begin{array}{c} \dfrac{\partial a_3}{\partial x_2} - \dfrac{\partial a_2}{\partial x_3} \\[2ex] \dfrac{\partial a_1}{\partial x_3} - \dfrac{\partial a_3}{\partial x_1} \\[2ex] \dfrac{\partial a_2}{\partial x_1} - \dfrac{\partial a_1}{\partial x_2} \end{array} \right\} \tag{D.19}
$$

Example D.11 (Curl of the velocity field). It can be shown that the curl of the velocity equals twice the angular velocity of rotation of the fluid particles about their center.

Definition D.10 (Laplacian). *The Laplacian of a scalar field* Φ *equals the scalar* $\Delta \Phi = \nabla \cdot \nabla \Phi$*, that is,*

$$
\nabla \cdot \nabla \Phi = \Delta \Phi = \frac{\partial^2 \Phi}{\partial x_1^2} + \frac{\partial^2 \Phi}{\partial x_2^2} + \frac{\partial^2 \Phi}{\partial x_3^2} \tag{D.20}
$$

The Laplacian of a fluid variable is related to transport phenomena by diffusion.

Example D.12 (Velocity potential). Under certain conditions (steady, ideal, incompressible and irrotational flow), the fluid field can be extracted from the gradient of a scalar field Φ, called *velocity potential*, so that

$$
\boldsymbol{v} = \nabla \Phi
$$

For a uniform flow, parallel to the x-axis, the velocity potential is $\Phi = V_\infty x$, which gives the two-dimensional velocity

$$
\boldsymbol{v} = \left\{ \begin{array}{c} \dfrac{\partial \Phi}{\partial x} \\[2ex] \dfrac{\partial \Phi}{\partial y} \end{array} \right\} = \left\{ \begin{array}{c} V_\infty \\ 0 \end{array} \right\}
$$

For this flow, $\nabla \cdot \boldsymbol{v} = 0$ and $\nabla \times \boldsymbol{v} = 0$, as expected.

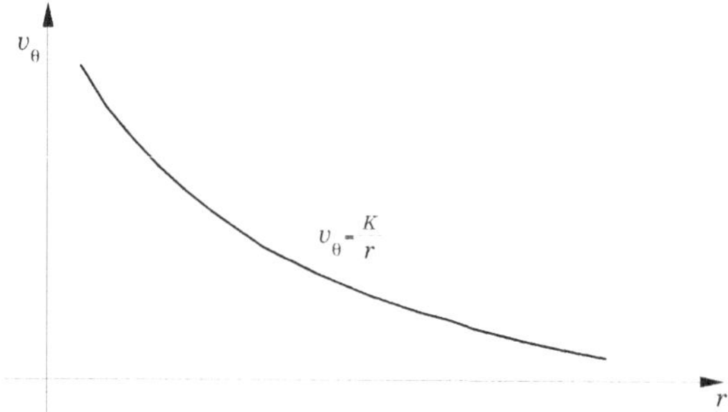

Problem D.20. Tangential velocity in a potential vortex.

Example D.13 (Potential vortex). In polar coordinates a potential vortex has the following velocity components

$$v_r = 0$$
$$v_\theta = \frac{K}{r}$$

with K a constant. In Cartesian coordinates,

$$v_x = -v_\theta \sin\theta = -K\frac{y}{x^2 + y^2}$$
$$v_y = v_\theta \cos\theta = K\frac{x}{x^2 + y^2}$$

As a consequence,

$$\operatorname{div} \boldsymbol{v} = \boldsymbol{\nabla} \cdot \boldsymbol{v} = \frac{\partial v_x}{\partial x} + \frac{\partial v_y}{\partial y} = K\left(-\frac{-2yx}{(x^2 + y^2)^2} + \frac{-2xy}{(x^2 + y^2)^2}\right) = 0$$

and the flow is *incompressible*. For a plane, two-dimensional flow, the curl has only the z component,

$$\operatorname{curl}_z \boldsymbol{v} = \frac{\partial v_y}{\partial x} - \frac{\partial v_x}{\partial y} = K\left(\frac{-2x^2}{(x^2 + y^2)^2} + \frac{1}{x^2 + y^2} + \frac{-2y^2}{(x^2 + y^2)^2} + \frac{1}{x^2 + y^2}\right) = 0$$

That is, the potential vortex is *irrotational*.

Example D.14 (Rigid body rotation). In polar coordinates, the rotation of the fluid as a rigid body corresponds to

$$v_r = 0$$

$$v_\theta = \omega r$$

Transforming into Cartesian coordinates,

$$v_x = -v_r \sin \theta = -\omega y$$

$$v_y = v_r \cos \theta = \omega x$$

As a consequence,

$$\operatorname{div} \boldsymbol{v} = \nabla \cdot \boldsymbol{v} = \frac{\partial v_x}{\partial x} + \frac{\partial v_y}{\partial y} = 0 + 0 = 0$$

and the flow is *incompressible*. For a plane, two-dimensional flow, the curl has only the z component,

$$\operatorname{rot}_z \boldsymbol{v} = \frac{\partial v_y}{\partial x} - \frac{\partial v_x}{\partial y} = \omega - (-\omega) = 2\omega$$

Therefore, the rotation of the flow as a rigid body is a *rotational* flow, where the curl equals twice the angular velocity of the fluid particle.

Problems

D.1   Given the velocity vector,

$$\boldsymbol{v} = \left\{ \begin{array}{c} 2x \\ 3x^2 \\ z^2 \end{array} \right\}$$

calculate

(a) $\dfrac{\partial v_i}{\partial x_i}$

(b) $\dfrac{\partial v_i}{\partial x_j}$

(c) Is the flow steady or unsteady?

(d) Is the flow compressible or incompressible?

D.2   The pressure field in a fluid flow is given by

$$p = p_\infty + \rho_\infty v_\infty^2 \left(\sin \frac{x}{a} \, \sin \frac{y}{b} + 2 \, \frac{x}{a} \right)$$

Find ∇p and $\varDelta p$ at the point (a, b) if p_∞, v_∞, a and b are constants. The pressure gradient is related to the net pressure-force over the fluid particle.

D.3 A scalar field is given by

$$\Phi(x, y) = 3x^2 y + 4y^2$$

Calculate $\nabla\Phi$ at $(3, 5)$. Find the component of $\nabla\Phi$ that forms an angle of $60°$ with the x axis at the point $(3, 5)$. (That is, find the projection of $\nabla\Phi$ on an axis that forms $60°$ with the x axis).

D.4 Consider the vectors c and d

$$\begin{aligned} c &= xyz\boldsymbol{i} + 2\boldsymbol{j} + y^2\boldsymbol{k} \\ d &= x^2\boldsymbol{i} + y^2\boldsymbol{j} + x\boldsymbol{k} \end{aligned}$$

and the scalar field

$$\phi = xy$$

Calculate (a)$c \cdot d$; (b)$c \times d$; (c)$\frac{\partial c}{\partial x}$; (d)$\nabla\phi$; (e)$\nabla \times c$; (f)$\nabla \cdot d$.

E

Useful Tools of Calculus

E.1 Taylor Expansion Series

In order to derive differential equations, many times one has to resort to the application of Taylor series. Taylor series relate the value of a function at a point x with the value and derivatives of the function at another point x_0.

Theorem E.1. *Let $f(x)$ be continuous and have continuous derivatives up to order $n + 1$ in the interval $(x_0 - a, x_0 + a)$. Then for each point x in this interval,*

$$
\begin{aligned}
f(x) \;=\; & f(x_0) + \frac{\mathrm{d}f(x_0)}{\mathrm{d}x}(x - x_0) + \frac{1}{2}\frac{\mathrm{d}^2 f(x_0)}{\mathrm{d}x^2}(x - x_0)^2 \\
& + \frac{1}{6}\frac{\mathrm{d}^3 f(x_0)}{\mathrm{d}x^3}(x - x_0)^3 + \cdots + \frac{1}{n!}\frac{\mathrm{d}^n f(x_0)}{\mathrm{d}x^n}(x - x_0)^n \\
& + R(x)
\end{aligned}
\tag{E.1}
$$

where $R(x)$ is the remainder or truncation error,

$$
R(x) = \frac{1}{(n+1)!}\frac{\mathrm{d}^{n+1} f(x_1)}{\mathrm{d}x^{n+1}}(x - x_0)^{n+1}
\tag{E.2}
$$

with x_1 a point in the interval (x_0, x) for $x > x_0$ or (x, x_0) for $x < x_0$.

Note that in order to derive differential equations it is enough to use the order of the truncation error, which can be approximated by

$$
R(x) \approx (x - x_0)^{n+1}
\tag{E.3}
$$

E.2 Gauss or Divergence Theorem

To manipulate the transport equations in integral form, we will employ the Gauss or divergence theorem. This theorem relates the domain integral of a derivative of a function to the boundary integral of a component of its flux.

Theorem E.2. *Let f be a continuous and differentiable function in the open domain Ω with boundary Γ and exterior normal $\boldsymbol{n}$. Then,*

$$\int_{\Omega} \frac{\partial f}{\partial x_i}\,\mathrm{d}\Omega = \int_{\Gamma} f n_i\,\mathrm{d}\Gamma \tag{E.4}$$

Corollary. Let $\boldsymbol{g}$ be a continuous and differentiable vector field in the open domain Ω with boundary Γ and exterior normal $\boldsymbol{n}$. Then,

$$\int_{\Omega} \nabla \cdot \boldsymbol{g}\,\mathrm{d}\Omega = \int_{\Gamma} \boldsymbol{g} \cdot \boldsymbol{n}\,\mathrm{d}\Gamma \tag{E.5}$$

F

Coordinate Systems

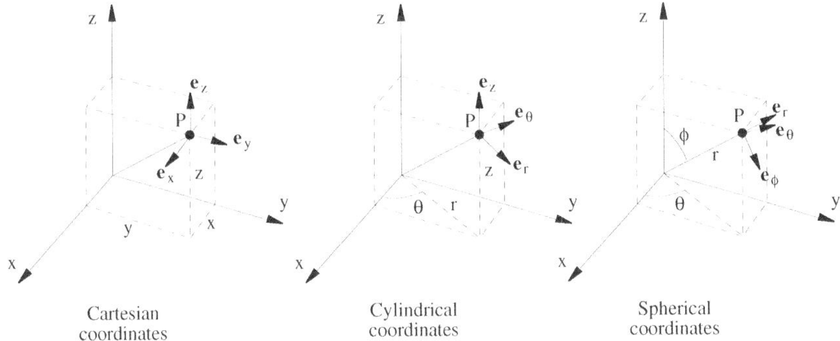

Fig. F.1. Coordinates of the point P for various orthogonal coordinate systems.

There are three main systems of orthogonal coordinates: Cartesian coordinates, cylindrical coordinates (called polar coordinates in 2D) and spherical coordinates. These are represented in Fig. F.1.

F.1 Cartesian Coordinates

They are frequently denoted by (x, y, z) or (x_1, x_2, x_3). The unit vectors in the coordinate axis are, respectively, i, j, k or e_x, e_y, e_z.

F.2 Cylindrical Coordinates

In this system of coordinates, the position of a point is given by the radius r with respect to the axis z, the angle θ to position the radius with respect to a reference direction and the height z. Thus, the coordinates are given by the set (r, θ, z) and the unit vectors along the axis by $\boldsymbol{e}_r$, $\boldsymbol{e}_\theta$, $\boldsymbol{e}_z$.

Note that in this coordinate system, the unit vectors $\boldsymbol{e}_r$ and $\boldsymbol{e}_\theta$ change with the point in space.

As a particular case in two dimensions we can find *polar coordinates*, similar to cylindrical coordinates, but without the z axis. Therefore, polar coordinates constitute a plane coordinate system.

Cylindrical (and polar) and Cartesian coordinates are related by

$$
\begin{aligned}
x &= r \cos \theta \\
y &= r \sin \theta \\
z &= z
\end{aligned}
\tag{F.1}
$$

whereas the inverse transformation is

$$
\begin{aligned}
r &= \sqrt{x^2 + y^2} \\
\theta &= \arctan \tfrac{y}{x} \\
z &= z
\end{aligned}
\tag{F.2}
$$

F.3 Spherical Coordinates

The spherical coordinates of a spatial point are given by the set (r, θ, ϕ) and the unit vectors along the axis by $\boldsymbol{e}_r$, $\boldsymbol{e}_\theta$, $\boldsymbol{e}_\phi$. The coordinate r is the radius of the point with respect to the origin of coordinates. The angles θ and ϕ are used to direct the radius in space (see Fig. F.1).

As for cylindrical coordinates, all the unit vectors vary with the position of the point.

Spherical and Cartesian coordinates are related by

$$
\begin{aligned}
x &= r \sin \phi \cos \theta \\
y &= r \sin \phi \sin \theta \\
z &= r \cos \phi
\end{aligned}
\tag{F.3}
$$

The inverse transformation is given by

$$
\begin{aligned}
r &= \sqrt{x^2 + y^2 + z^2} \\
\theta &= \arctan \tfrac{y}{x} \\
\phi &= \arctan \tfrac{\sqrt{x^2 + y^2}}{z}
\end{aligned}
\tag{F.4}
$$

G

Reference Systems

G.1 Definitions

Some fluid variables, like velocity, depend on the observer. The position and motion of the observer constitutes the system of reference. Systems of reference can be classified as inertial and non-inertial.

Definition G.1 (Inertial system of reference). *A system of reference is inertial, Galilean or absolute when it moves at a constant velocity. Therefore, there is no rotation nor acceleration of the observer. The case of a fixed (motionless) reference belongs to this type of reference. The fluid velocity with respect to this reference system is called absolute velocity.*

Definition G.2 (Non-inertial system of reference). *A system of reference is non-inertial when it is subjected to any acceleration (linear, angular or centripetal). The fluid velocity with respect to this system of reference is called relative velocity.*

G.2 Velocity Triangle

The velocity triangle is an important tool when relative systems of reference are employed. These find application in the analysis of moving or rotating objects, like the runner of a pump. In particular, the triangle of velocities relates the velocity vector expressed in inertial and non-inertial references.

Consider an absolute and a relative system of reference. The relative system moves at a velocity of v^{ref} with respect to the absolute system. Then, the absolute fluid velocity v^{abs} is related to the relative fluid velocity v^{rel} by the vector expression

$$\boxed{v^{\text{abs}} = v^{\text{rel}} + v^{\text{ref}}} \tag{G.1}$$

which is frequently called a *velocity triangle*.

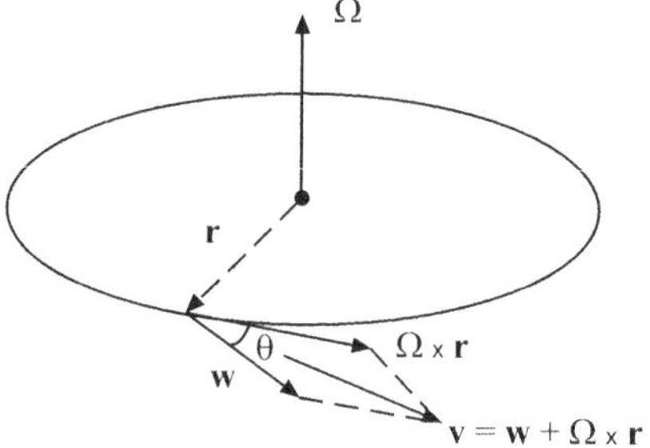

Fig. G.1. Composition of velocities in a rotating system of reference.

Example G.1 (Rotating system of reference). Consider a (non-inertial) rotating system of reference with angular velocity $\boldsymbol{\Omega} = \Omega\,\boldsymbol{e}_z$. If the relative velocity of a particle with respect to this system of reference in cylindrical coordinates is

$$
\boldsymbol{w} = \left\{ \begin{array}{c} w_r \\ w_\theta \\ w_z \end{array} \right\} = w \left\{ \begin{array}{c} \cos\theta \\ \sin\theta \\ 0 \end{array} \right\}
$$

determine the particle absolute velocity $\boldsymbol{v}$.

Solution. In cylindrical coordinates, the velocity of the system of reference is

$$
\boldsymbol{v}^{\mathrm{ref}} = \left\{ \begin{array}{c} v_r^{\mathrm{ref}} \\ v_\theta^{\mathrm{ref}} \\ v_z^{\mathrm{ref}} \end{array} \right\} = \boldsymbol{\Omega} \times \boldsymbol{r}
$$

$$
= \left\{ \begin{array}{c} 0 \\ 0 \\ \Omega \end{array} \right\} \times \left\{ \begin{array}{c} r \\ 0 \\ 0 \end{array} \right\}
$$

$$
= \left\{ \begin{array}{c} 0 \\ \Omega\,r \\ 0 \end{array} \right\}
$$

Using the velocity triangle,

$$
\boldsymbol{v}^{\mathrm{abs}} = \boldsymbol{v}^{\mathrm{rel}} + \boldsymbol{v}^{\mathrm{ref}}
$$

$$
= w \left\{ \begin{array}{c} \cos\theta \\ \sin\theta \\ 0 \end{array} \right\} + \left\{ \begin{array}{c} 0 \\ \Omega\,r \\ 0 \end{array} \right\}
$$

$$
= \left\{ \begin{array}{c} w\,\cos\theta \\ w\,\sin\theta + \Omega\,r \\ 0 \end{array} \right\}
$$

G.3 Conservation Equations for Non-Inertial Systems of Reference

The transport equations presented in Chapters 6 and 8 are applicable in principle to inertial systems of reference, where the fluid velocities are absolute velocities.

But they can also be applied to relative systems of reference. In this case v denotes the relative velocity with respect to the non-inertial system of reference v^{rel} and the body force needs to account for the inertial forces given by the acceleration of the system of reference. In doing so, we are indirectly accounting for the real particle acceleration. In particular,

$$f_m^{\mathrm{rel}} = f_m - (a_0 + \dot{\Omega} \times r + \Omega \times (\Omega \times r) + 2\Omega \times v) \tag{G.2}$$

where a_0 is the acceleration of the origin of the relative system, Ω is the angular velocity of the reference system, $\dot{\Omega}$, its angular acceleration, r the position vector and v the fluid velocity with respect to the relative system of reference.

Problems

G.1   Repeat the example of the rotating frame of reference for a position vector with a non-vanishing coordinate z.

H

Equations of State

In this Appendix a few concepts and examples of equations of state are reviewed.

H.1 Introduction

Definition H.1 (Equation of state). *Equations of state are functions that relate thermodynamic variables.*

For a *simple compressible substance* (a pure substance with only the reversible work mode of compression), the thermodynamic state is determined by two independent thermodynamic variables. For instance,

$$\rho = \rho(p, T)$$
$$e = e(p, T)$$

For *multicomponent systems* or systems with additional reversible modes of work, the state is determined by more variables. For example, the density of sea water with dissolved salt is a function of salt concentration, temperature and pressure.

Enthalpy

A very important equation of state is that of enthalpy. The specific enthalpy is defined as

$$h = e + \frac{p}{\rho} \tag{H.1}$$

where e is the specific internal energy, p the pressure and ρ the density. Very frequently, this equation is expressed as a function of the specific volume $v = 1/\rho$ instead of the density.

Entropy

The definition of the entropy depends on the number of reversible modes of work and the number of components in the mixture.

H.2 Simple Compressible Substance

For a simple compressible substance the thermodynamic state is determined by two independent thermodynamic variables, like p-T or ρ-T.

The specific entropy s can be calculated from the Gibbs relation,

$$
\begin{aligned}
T\mathrm{d}s \ &= \mathrm{d}e + p\mathrm{d}\frac{1}{\rho} \\
&= \mathrm{d}h - \frac{1}{\rho}\mathrm{d}p
\end{aligned}
\tag{H.2}
$$

The specific heats at constant volume and constant pressure are defined, respectively, as

$$
\begin{aligned}
c_v \ &= \ \left(\frac{\partial e}{\partial T}\right)_\rho \\
c_p \ &= \ \left(\frac{\partial h}{\partial T}\right)_p
\end{aligned}
\tag{H.3}
$$

Another important parameter is the specific heat ratio,

$$
\gamma = \frac{c_p}{c_v}
\tag{H.4}
$$

For air, $\gamma = 1.4$.

Thermally Perfect Substance

For a thermally perfect substance, the specific internal energy e is only a function of the temperature,

$$
e(p, T) = e(T)
$$

For a thermally perfect gas,

$$
h(p, T) = h(T)
$$

But note that for a thermally perfect liquid, $h(p, T) \neq h(T)$ (see remark H.1).

Calorically Perfect Substance

For a calorically perfect substance,

$$
\mathrm{d}e(T) = c_v \mathrm{d}T
$$

with c_v, the specific heat at constant volume, a constant. For a calorically perfect gas,

$$
\mathrm{d}h(T) = c_p \mathrm{d}T
$$

where the specific heat at constant pressure is constant.

Perfect Gas

The equation of state of a perfect gas can be given by

$$p = \rho R_{\text{gas}} T$$

where p is the thermodynamic pressure, ρ the density, T the temperature and R_{gas} the gas constant, which can be calculated as

$$R_{\text{gas}} = \frac{\mathcal{R}}{M_{\text{gas}}}$$

with $\mathcal{R} = 8\,314.36$ J/(kmol K) the universal gas constant in the SI system and M_{gas} the molar mass of the gas.

For a perfect gas,

$$R_{\text{gas}} = c_p - c_v \tag{H.5}$$

from where, using the specific heat ratio γ,

$$
\begin{aligned}
c_v &= \frac{R_{\text{gas}}}{\gamma - 1} \\
c_p &= \frac{\gamma R_{\text{gas}}}{\gamma - 1}
\end{aligned}
\tag{H.6}
$$

Integrating the Gibbs equation, the entropy of a perfect gas can be expressed as

$$s_2 - s_1 = c_p \ln \frac{T_2}{T_1} - R_{\text{gas}} \ln \frac{p_2}{p_1} \tag{H.7}$$

Liquid

The equation of state for a liquid is simply

$$\rho = \text{const}$$

For a liquid, the specific heats are equal and $\gamma = 1$.

Even for a calorically perfect liquid, the enthalpy depends on the pressure,

$$h = c_v(T - T_0) + e_0 + p/\rho \tag{H.8}$$

where T_0 is a reference temperature for the internal energy e_0.

Remark H.1. For a liquid, the pressure is not a thermodynamic variable, but a mechanical variable. Therefore, the enthalpy is a combination of thermodynamic and mechanical variables [20]. This is especially important when deriving transport equations for thermodynamic variables of liquids.

From the Gibbs equation, since the density is constant $d\rho = 0$,

$$\begin{aligned} T\mathrm{d}s \;&=\; \mathrm{d}e \\ &=\; c_v(T)\,\mathrm{d}T \end{aligned} \qquad\text{(H.9)}$$

For a calorically perfect liquid, the specific entropy can be written as

$$s_2 - s_1 = c_v \ln\frac{T_2}{T_1} \qquad\text{(H.10)}$$

H.3 Mixtures of Independent Substances

Let us assume that we have a mixture of n_{comp} constituents. If there are various phases, each chemical species at each phase may constitute a composition variable and, therefore, a constituent.

If the state of a constituent in the mixture can be evaluated independently of the other constituents, then the mixture is said to be of independent substances. For such a mixture and for one reversible mode of work, the thermodynamic state depends on two independent thermodynamic variables and the composition.

For such a mixture, all the constituents are at the same temperature T, and the internal energy, enthalpy, entropy and pressure can be calculated from the sum of those of the constituents evaluated at T and the corresponding partial pressure p_A,

$$\begin{aligned} e \;&=\; \sum_{A=1}^{n_{\mathrm{comp}}} Y_A\, e_A(T,p_A) \\ h \;&=\; \sum_{A=1}^{n_{\mathrm{comp}}} Y_A\, h_A(T,p_A) \\ s \;&=\; \sum_{A=1}^{n_{\mathrm{comp}}} Y_A\, s_A(T,p_A) \\ p \;&=\; \sum_{A=1}^{n_{\mathrm{comp}}} p_A \end{aligned} \qquad\text{(H.11)}$$

The Gibbs function may be written as

$$T\mathrm{d}s \;=\; \mathrm{d}e + p\,\mathrm{d}\frac{1}{\rho} + \sum_{A=1}^{n_{\mathrm{comp}}} \mu_A^{\mathrm{chem}}\,\mathrm{d}Y_A \qquad\text{(H.12)}$$

with μ_A^{chem} the specific electrochemical potential.

Note that in the presence of chemical reactions and interfaces, not all the compositions may be independent variables.

Mixture of Perfect Gases

For a mixture of n_{comp} perfect gases, the partial pressure is the pressure that each constituent would have if it occupied the whole volume,

$$p_A = \rho Y_A \frac{\mathcal{R}}{M_A} T \tag{H.13}$$

Indeed,

$$\begin{aligned}
\sum_{A=1}^{n_{\text{comp}}} p_A &= \sum_{A=1}^{n_{\text{comp}}} \rho Y_A \frac{\mathcal{R}}{M_A} T \\
&= \rho \left(\sum_{A=1}^{n_{\text{comp}}} \frac{\mathcal{R}}{M_A/Y_A} \right) T \\
&= \rho \left(\sum_{A=1}^{n_{\text{comp}}} \frac{\mathcal{R}}{M} \right) T \\
&= p
\end{aligned} \tag{H.14}$$

Since $Y_A = X_A M_A/M$, the above equation can be written as

$$p_A = X_A p \tag{H.15}$$

which is known as the *law of Dalton*.

Furthermore, the mixture's specific heats are

$$\begin{aligned}
c_v &= \sum_{A=1}^{n_{\text{comp}}} Y_A \, c_{vA} \\
c_p &= \sum_{A=1}^{n_{\text{comp}}} Y_A \, c_{pA}
\end{aligned} \tag{H.16}$$

The entropy of a mixture of perfect gases depends on the initial and final states of the components. For an isothermal and isobaric mixing process, the entropy change can be expressed as

$$s_2 - s_1 = - \sum_{A=1}^{n_{\text{comp}}} Y_A R_A \ln X_A \tag{H.17}$$

and it is due to the entropy production involved in the mixing process.

I

Multicomponent Reacting Systems

In Chapters 6 and 8 the transport equations for multicomponent systems with the same body force for all the components have been presented. In this appendix, these equations are extended for the case when the chemical species may be subject to different body forces. For example, this can take place in fluids with electrically charged substances [28, 18].

I.1 Mass Conservation

Since the body force does not enter this equation, it remains unmodified. For a control volume,

$$\frac{\mathrm{d}}{\mathrm{d}t} \int_{V_c(t)} \rho \; \mathrm{d}V + \int_{S_c(t)} \rho \left[(\boldsymbol{v} - \boldsymbol{v}^c) \cdot \boldsymbol{n} \right] \mathrm{d}S = 0 \tag{I.1}$$

The differential counterpart reads

$$\frac{\partial \rho}{\partial t} + \nabla \cdot (\rho \, \boldsymbol{v}) = 0 \tag{I.2}$$

I.2 Momentum Equation

Assume that the species A is subject to a body force $\boldsymbol{f}_{mA}$ $[\mathrm{m}^2/\mathrm{s}]$. Take an infinitesimal fluid volume $\mathrm{d}V$. The mass of species A inside this volume is $\rho_A \mathrm{d}V$ and, therefore, the force acting on that species is (see Chapter 3)

$$\mathrm{d}\boldsymbol{F}_{vA} = \boldsymbol{f}_{mA} \, \rho_A \mathrm{d}V \tag{I.3}$$

Then, the total force over the infinitesimal volume can be written as

$$\mathrm{d}\boldsymbol{F}_v = \sum_{A=1}^{n_{\mathrm{esp}}} \rho_A \boldsymbol{f}_{mA} \,\mathrm{d}V \tag{I.4}$$

which, integrated over a fluid volume, gives

$$\boldsymbol{F}_v = \sum_{A=1}^{n_{\mathrm{esp}}} \int_{V_f(t)} \rho_A \boldsymbol{f}_{mA} \,\mathrm{d}V \tag{I.5}$$

Thus, the integral equation for a control volume is

$$\frac{\mathrm{d}}{\mathrm{d}t} \int_{V_c(t)} \rho\boldsymbol{v} \,\mathrm{d}V + \int_{S_c(t)} \rho\boldsymbol{v} \left[(\boldsymbol{v} - \boldsymbol{v}^c) \cdot \boldsymbol{n} \right] \mathrm{d}S = \int_{S_c(t)} \boldsymbol{\tau}\boldsymbol{n} \,\mathrm{d}S \\ + \sum_{A=1}^{n_{\mathrm{esp}}} \int_{V_c(t)} \rho_A \boldsymbol{f}_{mA} \,\mathrm{d}V \tag{I.6}$$

and the differential form, substituting $\rho_A = \rho Y_A$,

$$\frac{\partial \rho\boldsymbol{v}}{\partial t} + \nabla \cdot (\rho\boldsymbol{v}\boldsymbol{v}) = -\nabla p + \nabla \cdot \boldsymbol{\tau}' + \rho \sum_{A=1}^{n_{\mathrm{esp}}} Y_A \boldsymbol{f}_{mA} \tag{I.7}$$

I.3 Total Energy Conservation

For this equation, we have to reformulate the power generated by the body forces. Taking into account that the species A moves at the speed $\boldsymbol{v}_A$, the power transmitted to the species A in the fluid volume $\mathrm{d}V$ can be written as

$$\mathrm{d}\dot{W}_{vA} = \mathrm{d}\boldsymbol{F}_{vA} \cdot \boldsymbol{v}_A = \rho_A \boldsymbol{f}_{mA} \cdot \boldsymbol{v}_A \,\mathrm{d}V \tag{I.8}$$

The power over the fluid volume $\mathrm{d}V$ will be

$$\mathrm{d}\dot{W}_v = \sum_{A=1}^{n_{\mathrm{esp}}} \rho_A \boldsymbol{f}_{mA} \cdot \boldsymbol{v}_A \,\mathrm{d}V \tag{I.9}$$

and integrating in the fluid volume

$$\dot{W}_v = \sum_{A=1}^{n_{\mathrm{esp}}} \int_{V_f(t)} \rho_A \boldsymbol{f}_{mA} \cdot \boldsymbol{v}_A \,\mathrm{d}V \tag{I.10}$$

Thus the total energy integral and differential equations become, respectively,

$$\frac{\mathrm{d}}{\mathrm{d}t} \int_{V_c(t)} \rho \left(e + \frac{1}{2} v^2 \right) \mathrm{d}V + \int_{S_c(t)} \rho \left(e + \frac{1}{2} v^2 \right) [(\boldsymbol{v} - \boldsymbol{v}^c) \cdot \boldsymbol{n}] \, \mathrm{d}S$$
$$= \int_{S_c(t)} (-p\boldsymbol{n} + \boldsymbol{\tau}'\boldsymbol{n}) \cdot \boldsymbol{v} \, \mathrm{d}S$$
$$+ \sum_{A=1}^{n_{esp}} \int_{V_c(t)} \rho_A \boldsymbol{f}_{mA} \cdot \boldsymbol{v}_A \, \mathrm{d}V$$
$$- \int_{S_c(t)} \boldsymbol{q} \cdot \boldsymbol{n} \, \mathrm{d}S + \int_{V_c(t)} \dot{q}_v \, \mathrm{d}V$$

$$\text{(I.11)}$$

and

$$\frac{\partial \rho(e + \frac{1}{2} v^2)}{\partial t} + \nabla \cdot \left(\rho(e + \frac{1}{2} v^2)\boldsymbol{v} \right) = \nabla \cdot (\boldsymbol{\tau} \boldsymbol{v}) + \rho \sum_{A=1}^{n_{esp}} Y_A \boldsymbol{f}_{mA} \cdot \boldsymbol{v}_A$$
$$- \nabla \cdot \boldsymbol{q} + \dot{q}_v$$

$$\text{(I.12)}$$

Recall that, as a function of the mass fluxes, the species velocity can be expressed as

$$\boldsymbol{v}_A = \boldsymbol{v} + \boldsymbol{j}_A/\rho_A \tag{I.13}$$

I.3.1 Mechanical Energy Equation

The mechanical energy equation is recovered with the dot product of the velocity times the momentum equation. The result is

$$\frac{\mathrm{d}}{\mathrm{d}t} \int_{V_c(t)} \rho \frac{1}{2} v^2 \, \mathrm{d}V + \int_{S_c(t)} \rho \frac{1}{2} v^2 [(\boldsymbol{v} - \boldsymbol{v}^c) \cdot \boldsymbol{n}] \, \mathrm{d}S$$
$$= \int_{S_c(t)} (-p\boldsymbol{n} + \boldsymbol{\tau}' \cdot \boldsymbol{n}) \cdot \boldsymbol{v} \, \mathrm{d}S + \int_{V_c(t)} p \nabla \cdot \boldsymbol{v} \, \mathrm{d}V$$
$$- \int_{V_c(t)} \phi_v \, \mathrm{d}V + \sum_{A=1}^{n_{esp}} \int_{V_c(t)} \rho_A \boldsymbol{f}_{mA} \cdot \boldsymbol{v} \, \mathrm{d}V$$

$$\text{(I.14)}$$

In differential form,

$$\frac{\partial \rho \frac{1}{2} v^2}{\partial t} + \nabla \cdot \left(\rho \frac{1}{2} v^2 \boldsymbol{v} \right) = \nabla \cdot (\boldsymbol{\tau} \boldsymbol{v}) + p \nabla \cdot \boldsymbol{v} - \phi_v + \rho \boldsymbol{v} \cdot \sum_{A=1}^{n_{esp}} Y_A \boldsymbol{f}_{mA} \tag{I.15}$$

I.3.2 Internal Energy Equation

The internal energy equation is obtained by subtracting the mechanical energy equation from the total energy equation. For a control volume, the result is

$$\frac{\mathrm{d}}{\mathrm{d}t} \int_{V_c(t)} \rho e \ \mathrm{d}V + \int_{S_c(t)} \rho e \left[(\boldsymbol{v} - \boldsymbol{v}^c) \cdot \boldsymbol{n} \right] \mathrm{d}S$$

$$= - \int_{S_c(t)} p \boldsymbol{\nabla} \cdot \boldsymbol{v} \ \mathrm{d}V + \int_{V_c(t)} \phi_v \ \mathrm{d}V - \int_{S_c(t)} \boldsymbol{q} \cdot \boldsymbol{n} \ \mathrm{d}S + \int_{V_c(t)} \dot{q}_v \ \mathrm{d}V$$

$$+ \sum_{A=1}^{n_{\mathrm{esp}}} \int_{V_c(t)} \boldsymbol{f}_{mA} \cdot \boldsymbol{j}_A \ \mathrm{d}V$$

$$(\mathrm{I}.16)$$

and the differential form,

$$\frac{\partial \rho e}{\partial t} + \nabla \cdot (\rho e \boldsymbol{v}) = -p \boldsymbol{\nabla} \cdot \boldsymbol{v} + \phi_v - \boldsymbol{\nabla} \cdot \boldsymbol{q} + \dot{q}_v + \sum_{A=1}^{n_{\mathrm{esp}}} \boldsymbol{f}_{mA} \cdot \boldsymbol{j}_A \qquad (\mathrm{I}.17)$$

I.4 Conservation of Chemical Species

This equation remains unmodified. Thus, the integral equation reads

$$\frac{\mathrm{d}}{\mathrm{d}t} \int_{V_c(t)} \rho_A \ \mathrm{d}V + \int_{S_c(t)} \rho_A \left[(\boldsymbol{v} - \boldsymbol{v}^c) \cdot \boldsymbol{n} \right] \mathrm{d}S = \ - \int_{S_c(t)} \rho_A (\boldsymbol{v}_A - \boldsymbol{v}) \cdot \boldsymbol{n} \ \mathrm{d}S$$

$$+ \int_{V_c(t)} \dot{\omega}_A \ \mathrm{d}V$$

$$(\mathrm{I}.18)$$

and the differential form,

$$\frac{\partial \rho_A}{\partial t} + \nabla \cdot (\rho_A \ \boldsymbol{v}) = -\nabla \cdot \boldsymbol{j}_A + \dot{\omega}_A \qquad (\mathrm{I}.19)$$

I.5 Generalized Fourier's and Fick's laws

The laws of Fourier and Fick presented in Chapter 7 are simplified versions of the complete constitutive equations. Thermal and mass diffusion can be caused by more physical phenomena. For instance, concentration gradients can induce heat transport and, vice versa, temperature gradients can cause mass transport. For further details see [12, 3, 28, 18].

I.5.1 Heat Transport

Heat transport by diffusion can be caused by three effects: temperature gradients, mass diffusion and concentration gradients,

$$\boldsymbol{q} = \boldsymbol{q}^{\mathrm{T}} + \boldsymbol{q}^{\mathrm{m}} + \boldsymbol{q}^{\mathrm{Dufour}} \qquad (\mathrm{I}.20)$$

The first term, q^{T}, due to temperature gradients, is given by Fourier's conduction model

$$q^{\mathrm{T}} = -\kappa \nabla T \tag{I.21}$$

The second term, q^{m}, is the heat transport due to mass diffusion, and it is modeled as

$$q^{\mathrm{m}} = \sum_{A=1}^{n_{\mathrm{esp}}} h_A \boldsymbol{j}_A \tag{I.22}$$

where h_A is the specific enthalpy of the species A. This contribution stems from the introduction of the average fluid velocity $\boldsymbol{v}$, whereas each species presents its own velocity $\boldsymbol{v}_A$.

Finally, the *Dufour* or *diffusion-thermo* effect, which can be neglected in many engineering applications, is caused by differences of species velocity $\boldsymbol{v}_A$. Sometimes, it is thought of as thermal diffusion generated by concentration gradients. In the case of a multicomponent gas it can be modeled by [18]

$$q^{\mathrm{Dufour}} = \mathcal{R}T \sum_{A=1}^{n_{\mathrm{esp}}} \sum_{B=1}^{n_{\mathrm{esp}}} \frac{X_B D_A^T}{M_A D_{AB}} \left(\frac{\boldsymbol{j}_A}{\rho_A} - \frac{\boldsymbol{j}_B}{\rho_B} \right) \tag{I.23}$$

where $\mathcal{R}$ is the universal gas constant; D_A^T are the thermal diffusion coefficients; and D_{AB}, the binary diffusivities.

Radiation

Still, in high temperature applications, the heat transport by radiation may play an important role. This effect can be added to the constitutive equation of q by adding the term q^R, which is a highly nonlinear function of the temperature and the geometry of the problem.

I.5.2 Mass Transport

Mass transport by diffusion can be produced by concentration gradients, temperature gradients, pressure gradients and existence of different body forces for each chemical species,

$$\boldsymbol{j}_A = \boldsymbol{j}_A^{\mathrm{m}} + \boldsymbol{j}_A^{\mathrm{T}} + \boldsymbol{j}_A^{\mathrm{P}} + \boldsymbol{j}_A^{\mathrm{f_m}} \tag{I.24}$$

Typically, it is formulated as an implicit equation [18, 3] for $\boldsymbol{j}_A$,

$$
\begin{aligned}
\nabla X_A \;=\; & \sum_{B=1}^{n_{\mathrm{esp}}} \frac{X_A X_B}{D_{AB}} \left(\frac{\boldsymbol{j}_A}{\rho_A} - \frac{\boldsymbol{j}_B}{\rho_B} \right) \\[4pt]
+ \;& (Y_A - X_A)\, \nabla (\ln p) \\[4pt]
+ \;& \frac{\rho}{p} \sum_{B=1}^{n_{\mathrm{esp}}} Y_A Y_B\, (\boldsymbol{f}_{mA} - \boldsymbol{f}_{mB}) \\[4pt]
+ \;& \sum_{B=1}^{n_{\mathrm{esp}}} \frac{X_A X_B}{\rho D_{AB}} \left(\frac{D_B^T}{Y_B} - \frac{D_A^T}{Y_A} \right) \nabla (\ln T)
\end{aligned}
\tag{I.25}
$$

where D_A^T are the thermal diffusion coefficients; and D_{AB}, the binary diffusivities.

Temperature gradients can induce mass diffusion. This is called the *thermal diffusion* or *Soret* effect and is modeled as

$$\boldsymbol{j}_A^{\mathrm{T}} = -D_A^T \boldsymbol{\nabla}(\ln T) \tag{I.26}$$

where D_A^T is the *thermal diffusion* coefficient.

When the body force acting upon all the species is the same, diffusion by body forces cancels out.

Binary Systems

For binary systems $(1 \leq A, B \leq 2)$, mass transport can be simplified to [10]

$$\begin{aligned}
\boldsymbol{j}_A \quad &= -\rho D_{AB}\boldsymbol{\nabla}Y_A - D_A^T\boldsymbol{\nabla}(\ln T) + \frac{M_A M_B D_{AB} Y_A}{p}\,\boldsymbol{\nabla}p \\
&+ \frac{\rho M_A M_B D_{AB} Y_A Y_B}{p}\,(\boldsymbol{f}_{mA} - \boldsymbol{f}_{mB})
\end{aligned} \tag{I.27}$$

I.6 Chemical Production

In this section we hint how to calculate the species production term $\dot{\omega}_A$ that appears in the chemical species transport equation.

Assume that the reactions in a system can be modeled by K multistep *reversible* reactions, which are represented by

$$\sum_{i=1}^{n_{\mathrm{esp}}} \nu'_{A,k}\mathcal{M}_A \rightleftharpoons \sum_{i=1}^{n_{\mathrm{esp}}} \nu''_{A,k}\mathcal{M}_A \qquad k = 1, 2, \ldots, K \tag{I.28}$$

where $\mathcal{M}_A$ is the chemical symbol of species A, and $\nu_{A,k}$ the corresponding molar concentration coefficients.

The generalized law of mass action states that the characteristic molar production per unit volume per second for each reaction is modeled as

$$\hat{\omega}'_k = k_{k,f} \prod_{A=1}^{n_{\mathrm{esp}}} c_A^{\nu'_{A,k}} - k_{k,b} \prod_{A=1}^{n_{\mathrm{esp}}} c_A^{\nu''_{A,k}} \tag{I.29}$$

where $k_{k,f}$ and $k_{k,b}$ are the reaction rate for the forward and backward reactions, respectively. These are temperature dependent and can be modeled with the Arrhenius or modified Arrhenius law [18].

Encompassing all the equations, the molar production of species A is

$$\dot{\omega}'_A = \sum_{k=1}^{K}(\nu''_{A,k} - \nu'_{A,k})\hat{\omega}'_k \tag{I.30}$$

which is equivalent to the mass production per unit volume per unit time

$$\dot{\omega}_A = M_A \dot{\omega}'_A \tag{I.31}$$

Problem Solutions

Solutions for Chapter 1

1.1 $\sigma = 37.1$ MPa, $\tau = 1.8$ MPa

Solutions for Chapter 2

2.1 $a = (3, 3tz + txy^2, y^2 + 2xyzt)$

2.2 streamline: $\dfrac{y}{y_0} = \left(\dfrac{x}{x_0}\right)^n$, $n = \dfrac{1+t}{1+2t}$;

trajectory: $\dfrac{y}{y_0} = \left(2\dfrac{x}{x_0} - 1\right)^{1/2}$;

streakline: $\dfrac{y}{y_0} = \left(\dfrac{1+2t}{1+2\left((1+t)\frac{x_0}{x} - 1\right)}\right)^{1/2}$.

2.3 $r = C \exp(-\frac{a}{b}\theta)$

2.4 (a) $\dfrac{x(t)}{x_0} = \exp 5t(1 + t/2)$, $\dfrac{y(t)}{y_0} = \exp 5t(-1 + t/2)$. Eulerian description.

(b) $\left(\dfrac{x}{x_0}\right)^{-1+t} = \left(\dfrac{y}{y_0}\right)^{1+t}$

(c) $\dfrac{x}{x_0} = \dfrac{\exp 5t(1 + t/2)}{\exp 5\xi(1 + \xi/2)}$, $\dfrac{y}{y_0} = \dfrac{\exp 5t(-1 + t/2)}{\exp 5\xi(-1 + \xi/2)}$, $\xi < t$, where ξ is the equation parameter.

2.5 $xy = C$, $a = (a^2 x, a^2 y)$

2.6 $x^2 + y^2 = C$

2.7 $r = C$

2.8 $Q = \frac{4}{3} bw \sin 60^\circ \, v_0$, $\dot{m} = \frac{4}{3} \rho bw \sin 60^\circ \, v_0$

2.9 $\frac{(\rho_l - \rho_v) g \delta^3}{3\mu}$

2.10 $\int_S (\rho \frac{1}{2} v^2) \boldsymbol{v} \cdot \boldsymbol{n} \, dS$

Solutions for Chapter 3

3.1 5.23 kgf/cm^2

3.2 Smaller values of p_{gag} would imply negative absolute pressure.

3.3 $\sigma = \frac{F}{2\pi D}$

Solutions for Chapter 4

4.1 (a) $p = 49.05$ MPa, (b) $p = 49.66$ MPa

4.2 $\boldsymbol{F} = 294 \, \boldsymbol{i} - 509 \, \boldsymbol{j} \text{ kN}$

4.3 $T = \frac{1}{2} \rho g R_0^2$

4.4 $h = 1 + \sqrt{3} \cdot 0.2 \text{ m}$

4.5 $h = 0.33 \text{ m}$

4.6 (a) $M = 623.7$ kg
(b) $F_N = 0$

4.7 $\boldsymbol{F} = -p_{cg} \, A \, \boldsymbol{n}$, $h_{cp} = h_{cg} + \frac{\sin^2 \theta I_{yy}}{h_{cg} A}$

4.8 $p = \rho g l \sin \theta$

4.9 $\Delta p = (\rho_{\text{Hg}} - \rho_{\text{H}_2\text{O}}) g h$

Solutions for Chapter 6

6.1 (a) $V_s = V \frac{R}{2b}$; $Q = \pi R^2 V$
(b) $V_{\max} = V \frac{3R}{4b}$; $Q = \pi R^2 V$

6.2 $V_s = V \left(\frac{D}{d} \right)^2$; $Q = \frac{\pi D^2}{4} V = \frac{\pi d^2 V_s}{4}$

6.3 $\left(\dfrac{\partial \rho}{\partial t}\right)_{t=0} = -2.48 \text{ kg/m}^3\text{s}$

6.4 $\mathrm{d}M/\mathrm{d}t = -0.43 \text{ kg/s}$

6.5 $F_x = -370.5 \text{ N}$

6.6 $V = \sqrt{\dfrac{4W}{\rho \pi D_0^2}}$

6.7 $W = 2\,566 \text{ W}$

6.8 $F_x = 10^5 \text{ N}, \ F_y = -0.25 \tan\theta - 20\,400 \text{ N}$

6.9 (a) $P = \rho S_e (V - V_c)^2 (1 - \cos\theta) V_c$
(b) $P = \rho S_e V (V - V_c)(1 - \cos\theta) V_c$

6.10 $\theta = 0° \qquad \Omega = 43.3 \text{ rad/s}$
$\theta = 40° \qquad \Omega = 33.2 \text{ rad/s}$

6.11 $v_c = 1.84 \text{ m/s}, \ \rho_c = 953.8 \text{ kg/m}^3$

6.12 $v_c = 1.91 \text{ m/s}, \ \rho_c = 953.8 \text{ kg/m}^3$

6.13 $\left.\dfrac{\mathrm{d}h}{\mathrm{d}t}\right|_{h=1 \text{ m}} = 2.7 \times 10^{-3} \text{ m/s}, \ t = 82.1 \text{ s}$

6.14 $h_{pA} = 1.191 \text{ m}$

6.15 $p_3 = 71 \text{ atm}$

6.16 (a) $V_2 = A_1 V_1 / A_2$
(b) $\Delta p/\rho = V_1^2 \left(1 - \dfrac{A_1}{A_2}\right) \dfrac{A_1}{A_2}$
(c) $D_v = \frac{1}{2}\rho A_1 \left(1 - \dfrac{A_1}{A_2}\right)^2 V_1^3$
(d) $K_s - \left(1 - \dfrac{A_1}{A_2}\right)^2$

6.17 $\dot{Q} = M\,\omega$

6.18 $t = \dfrac{1}{5} \left(\dfrac{z_0}{z_2}\right)^2 \sqrt{\dfrac{2z_0}{g}}$

6.19 $\rho_s(t) = \rho_{s0}\, e^{-\frac{Q}{V}t}$

6.20 (a) Rate of consumption $-100\, c(t) \text{ mol/s}$
(b) $c(t) = e^{-0.1t} \text{ mol/l}$
(c) Heat generation $\dfrac{\mathrm{d}Q}{\mathrm{d}t} = 500\, c(t) \text{ cal/s}$

6.21 $T(t) = T_0 + \dfrac{5.0}{\rho c_p}(1 - \exp(-0.1t))$

6.22 (a) $u_3 = 1.79 \text{ m/s}$
(b) $c = 8 \times 10^{-2} \text{gr/cm}^3$
(c) $\dot{Q} = 48 \text{ W}$

Solutions for Chapter 7

7.1 $P = 2034$ W

7.2 4 times

7.3 $\kappa = 0.015$ W/(m K)

7.4 $F = 1.08$ N

7.5 $\delta = 0.026$ m

7.6 $V = \dfrac{mgh\,\sin\theta}{\mu A}$

Solutions for Chapter 9

9.1 $[F] = \mathsf{M}^1\mathsf{L}^1\mathsf{T}^{-2}$, $[\sigma] = \mathsf{M}^1\mathsf{L}^{-1}\mathsf{T}^{-2}$, $[\dot{W}] = \mathsf{M}^1\mathsf{L}^2\mathsf{T}^{-3}$, $[\mu] = \mathsf{M}^1\mathsf{L}^{-1}\mathsf{T}^{-1}$, $[\kappa] = \mathsf{M}^1\mathsf{L}^1\mathsf{T}^{-3}\Theta^{-1}$.

9.2 $\dfrac{F}{\rho V^2 L^2} = \phi(\dfrac{\rho V L}{\mu}, \dfrac{U^2}{gL})$

9.3 $\dfrac{P}{\rho D^5 \Omega^3} = f\left(\dfrac{Q}{D^3 \Omega}\right)$

9.4 3 parameters: $\dfrac{\sigma}{\rho D V^2}$, $\dfrac{\mu}{\rho V D}$, $\dfrac{d}{D}$

9.5 $\dfrac{T}{\mu R^3 w} = f\left(\dfrac{h}{R}\right)$

9.6 $\dfrac{Q}{V_o H^2} = f\left(\dfrac{V_o}{\sqrt{gH}}, \phi\right)$; $\dfrac{Q}{\sqrt{gH}H^2} = f'(\phi)$.

9.7 (a) 2
(b) No
(c) $F_p = F_m$, $V_m = 50$ m/s

9.8 $V = 9.58$ m/s

9.9 The nondimensional numbers must be maintained both for the model and prototype, $\dfrac{gh}{D^2 \Omega^2} = f\left(\dfrac{gH}{D^2 \Omega^2}, \dfrac{L}{D}\right)$

9.10 $\dfrac{D}{\rho V^2 h^2} = f\left(\dfrac{\rho V h}{\mu}, \dfrac{w}{h}\right)$

9.11 $\mathrm{Re} \approx 9 \times 10^6$

9.12 (a) Con $\dfrac{V_c}{D^{1/2}g^{1/2}} = f\left(\dfrac{D_p}{D}, \dfrac{\rho_p}{\rho}, \dfrac{\mu}{\rho D^{3/2}g^{1/2}}\right)$, $V_c = 1.414$ m/s

(b) Con $\dfrac{\rho V_c D}{\mu} = f\left(\dfrac{D_p}{D}, \dfrac{\rho_p}{\rho}, \dfrac{\rho^2 D^3 g}{\mu^2}\right)$, $V_c = 0.5$ m/s

(c) One is working under conditions of partial similarity and the results depend upon the dimensionless numbers chosen.

(d) The solution is to carry out the experiment under complete similarity conditions, with a liquid of kinematic viscosity $\nu_m = \nu_p/\sqrt{8}$. In this case $V_c = 1.414$ m/s

9.13 (a) $\dfrac{tU}{L} = \varphi\left(\dfrac{\delta}{L}, \dfrac{P_v}{\rho U^2}, \dfrac{\mu}{\rho U L}\right)$

(b) $\delta_m = 0.002$ m, $U_m = 10$ m/s, $P_{vm} = 50\,000$ Pa

(c) $t_p = 250$ min

Solutions for Chapter 10

10.1 (a) $[\epsilon] = \mathsf{L}^2\mathsf{T}^{-1}$, $[R] = \mathsf{M}\mathsf{L}^{-2}\mathsf{T}^{-1}$, $[\omega_f] = \mathsf{L}^1\mathsf{T}^{-1}$.

(b) $\Pi_t = \dfrac{l_0/t_0}{u_0}$, $\Pi_\epsilon = \dfrac{\epsilon_0}{u_0 l_0}$, $\Pi_R = \dfrac{R l_0}{u_0 h_0 S_0}$, $\Pi_\omega = \dfrac{\omega_f l_0}{u_0 h_0}$.

(c) $\Pi_t = \mathrm{S}$, $\Pi_\epsilon = 1/\mathrm{Pe_{II}}$, $\Pi_R = \mathrm{Da_I}$.

10.2 (a) $\Pi_t = \dfrac{l_0/t_0}{u_0} \ll 1$

(b) $\Pi_\epsilon = \dfrac{\epsilon_0}{u_0 l_0} \ll 1$

(c) $\Pi_\omega = \dfrac{\omega_f l_0}{u_0 h_0} \gg 1$

(d) It is steady flow, with small viscous forces and sedimentation of the order of convection.

10.3 $\mathrm{S} = \dfrac{l_0/v_0}{t_0}$, $\mathrm{Pe_{II}} = \dfrac{v_0 l_0}{\kappa_0}$

10.4 (a) $[\epsilon] = 1$

(b) $\Pi_t = \dfrac{\epsilon \mu l_0^2}{\gamma t_0 \Delta p_0}$, $\Pi_g = \dfrac{\rho_0 g l_0}{\Delta p_0}$

10.5 (a) $[\gamma] = \mathrm{m}^2$

(b) $\Pi = \dfrac{\kappa \mu}{\rho_0 c_p \gamma \Delta p_0}$ with $\Delta p_0 = \rho_0 v_0^2$

10.6 $\Pi_t = \dfrac{l_0/t_0}{v_0}$, $\Pi_{\Delta p} = \dfrac{\Delta p}{\rho_0 v_0^2}$, $\Pi_\mu = \dfrac{\mu}{\rho_0 v_0 l_0}$, $\Pi_g = \dfrac{g l_0}{v_0^2}$, $\Pi_\beta = \dfrac{\beta \Delta T_0 g l_0}{v_0^2}$.

10.7 (a) Centrifugal force: $\dfrac{\omega_0^2 l_0^2}{v_0^2}$, Coriolis force: $\dfrac{\omega_0 l_0}{v_0}$

(b) S, Eu, Re, Fr.

10.8 (a) $[\gamma] = \mathsf{L}^2$

(b) $\Pi_p = \dfrac{\gamma p_0}{\mu v_0 l_0}$, $\Pi_\gamma = \dfrac{\gamma}{l_0^2}$, $\Pi_g = \dfrac{\rho g \gamma}{\mu v_0}$

Solutions for Chapter 12

12.1 164 Pa

12.2 206 165 Pa

12.3 4.37 MW

12.4 2.0×10^{-6} kg/(m s)

12.5 41.5 °C

12.6 $\dfrac{\overline{h}_L}{h_x} = \dfrac{1}{0.9}\left(\dfrac{x}{L}\right)^{0.1}$

12.7 $h_m = 1.85 \times 10^{-3}$ m/s

12.8 $j_{CO_2} = -1.0 \times 10^{-6}$ kg/(m^2 s)

12.9 $h_m = \dfrac{D_{AB} E}{C_{A\infty} - F}$

12.10 $C_f = \dfrac{2\mu A}{\rho U_\infty^2}$, $h = \dfrac{\kappa E}{T_\infty - D}$

12.11 5.9 °C

12.12 (a) $J = 1.429 \times 10^{-3}$ kg/s
(b) $\dot{Q} = 3725$ W

12.13 $\mathrm{Nu} = \phi(\mathrm{Gr}, \mathrm{Ja}, \mathrm{Pr}, \mathrm{Bo})$, where $\mathrm{Gr} = \dfrac{\rho g(\rho_{\mathrm{liq}} - \rho_{\mathrm{vap}})L^3}{\mu^2}$, $\mathrm{Ja} = \dfrac{c_p \Delta T}{h_{lv}}$ (Jakob number) and $\mathrm{Bo} = \dfrac{(\rho_{\mathrm{liq}} - \rho_{\mathrm{vap}})gL^2}{\sigma}$ (Bond number).

Solutions for Chapter 13

13.1 (a) $\dot{Q} = \pi\mu\dfrac{\omega^2 d^3 L}{D - d}$
(b) $\dfrac{dT}{dt} = \dfrac{4\mu\omega^2 d^3}{\rho c_v (D - d)^2 (D + d)}$

13.2 (a) $\nabla \cdot \boldsymbol{v} = 3$
(b) Compressible, because $\nabla \cdot \boldsymbol{v} \neq 0$
(c) $\phi_v = \mu(100y^2 + 18) + \lambda 9$

13.3 (a) $\dfrac{P}{\rho D^5 \omega^3} = const$
(b) $\dfrac{L_m}{L_p} = \sqrt[3]{3}$
(c) $P_m = 3P_p$; $\omega_m = 3^{-2/9}\omega_p$

13.4 $F_x = -(p_1 \cos 30 A_1 + p_2 A_2) - \rho(V_1^2 A_1 \cos 30 + V_2^2 A_2)$
$F_y = \rho g V - (p_1 \sin 30 A_1)$

13.5 (a) $\dot{m} = \frac{3}{2}\rho U_0 dL$

(b) $F_D = \rho U_0^2 dL$; $C_f = \dfrac{2}{\pi}$

(c) $T_s = T_0 + \dfrac{3}{4} \dfrac{\dot{Q} + \frac{3}{4}\rho U_0^3 dL}{\rho c_v U_0 dL}$

13.6 $F_1 = \rho g (H - h)a$; $F_2 = \frac{1}{2}\rho g H^2$; $y_2 = \frac{H}{3} + (H - h)\left(\frac{a}{H}\right)^2$

13.7 (a) Dilatant
(b) $\tau_0' = \mu a_1^n$

13.8 (a) $h_m = 0.0188$ m/s
(b) $\dot{Q}_{\text{total}} = 51.81$ W

13.9 (a) If $\rho_A = \rho_B = \rho_{AB}$, $u_2^{AB} = 1.6$ m/s, $\Delta p = 240$ Pa.

$$\frac{\Delta p}{\rho_A (u_1^A)^2} = \left\{ 1 + \left(\frac{u_1^B}{u_1^A}\right)^2 - 2\,\frac{u_1^B}{u_1^A} \right\} \left(\frac{d}{D}\right)^2 \left[1 - \left(\frac{d}{D}\right)^2 \right]$$

(b) $$\frac{\Delta p}{\rho_A (u_1^A)^2} = \left\{ 1 + \frac{\rho_B}{\rho_A}\left(\frac{u_1^B}{u_1^A}\right)^2 - \left(1 + \frac{\rho_B}{\rho_A}\right)\frac{u_1^B}{u_1^A} \right\} \left(\frac{d}{D}\right)^2 \left[1 - \left(\frac{d}{D}\right)^2 \right]$$

(c) It can be observed that (a) is a particular case of (b). Furthermore, reorganizing (b) as

$$\frac{\Delta p}{\rho_A (u_1^A)^2} = \left\{ 1 + \frac{\rho_B}{\rho_A}\frac{u_1^B}{u_1^A}\left(\frac{u_1^B}{u_1^A} - 1\right) - \frac{u_1^B}{u_1^A} \right\} \left(\frac{d}{D}\right)^2 \left[1 - \left(\frac{d}{D}\right)^2 \right]$$

for $u_1^B > u_1^A$, if $\rho_B > \rho_A$ then $\Delta p > \Delta p_{\rho=\text{const}}$ and if $\rho_B < \rho_A$ then $\Delta p < \Delta p_{\rho=\text{const}}$. And for $u_1^A > u_1^B$, if $\rho_A > \rho_B$ then $\Delta p > \Delta p_{\rho=\text{const}}$ if if $\rho_A < \rho_B$ then $\Delta p < \Delta p_{\rho=\text{const}}$.

13.10 $M = \dfrac{\pi}{32}\dfrac{\mu \omega D^4}{h \sin \alpha}$

13.11 $Q = \dfrac{\rho g \pi R^4}{8\mu}$

13.12 $\dfrac{1}{\text{Re}}\dfrac{\mathrm{d}^4 \phi'(y')}{\mathrm{d}(y')^4} - \left[\omega' + \dfrac{2}{\text{Re}}(k')^2\right]\dfrac{\mathrm{d}^2 \phi'(y')}{\mathrm{d}(y')^2} + (k')^2\left[\omega' + \dfrac{1}{\text{Re}}(k')^2\right]\phi'(y') = 0$

13.13 (a) $t\sqrt{\dfrac{g}{d}} = f\left(\dfrac{D}{d}, \dfrac{h}{d}\right)$
(b) $h' = 4h$; $t' = 2t$
(c) The same
(d) $\Pi_\mu = \dfrac{\mu}{\rho d^{3/2} g^{1/2}}$; there will be complete similitude only if $\mu'/\rho' = \mu/\rho$.

13.14 $A = 7.9 \times 10^{-3}\,\mathrm{m}^2$

13.15 $\dfrac{y}{y_0} = e^{3\left(\frac{1}{x} - \frac{1}{x_0}\right)}$

13.16 $p(r) = p_{\mathrm{atm}} + \dfrac{2\pi K}{3}\left(R^2 - r^2\right)$

13.17 $t = 128, 45$ hours

13.18 (a) $\dfrac{gH}{N^2 D^2} = f\left(\dfrac{Q}{ND^3}, \mathrm{Re}, \dfrac{\epsilon}{D}\right)$

(b) $\dfrac{gH}{N^2 D^2} = f\left(\dfrac{Q}{ND^3}\right)$

(c) $H'(Q') = 80 - 0.1(Q')^2$

(d) The same.

13.19 (a) $Q_3 = Q_1 + Q_2$

(b) $D_v = \left[\frac{1}{2}\rho\left(\dfrac{Q_1^3}{S_1^2} + \dfrac{Q_2^3}{S_2^2}\right) + (p_1 Q_1 + p_2 Q_2)\right] - \left[\frac{1}{2}\rho\dfrac{Q_3^3}{S_3^2} + p_3 Q_3\right] + \dot{W}$

(c) $T_3 = \dfrac{T_1 Q_1 + T_2 Q_2}{T_3 Q_3} + \dfrac{D_v}{\rho c_v Q_3}$

13.20 (a) $\mathrm{Nu}_L = \frac{4}{3}C\left(\dfrac{\mathrm{Gr}_L}{4}\right)^{1/4}$

Solutions for Appendix D

D.1 (a) $2(1 + z)$

(b) $\begin{pmatrix} 2 & 0 & 0 \\ 6x & 0 & 0 \\ 0 & 0 & 2z \end{pmatrix}$

(c) Steady

(d) Compressible

D.2 $\nabla p = \rho_\infty v_\infty^2 \left\{ \dfrac{2}{a} + \dfrac{1}{a}\cos\dfrac{x}{a}\sin\dfrac{y}{b}, \dfrac{1}{b}\sin\dfrac{x}{a}\cos\dfrac{y}{b}\right\}$

$\Delta p = -\rho_\infty v_\infty^2 \left(\dfrac{1}{a^2} + \dfrac{1}{b^2}\right)\sin\dfrac{x}{a}\sin\dfrac{y}{b}$

D.3 $\nabla\phi = (90, 67)$; the projection is 103.02.

D.4 (a) $x^3 yz + 2y^2 + xy^2$

(b) $(2x - y^4, x^2 y^2 - x^2 yz, xy^3 z - 2x^2)$

(c) $yz\mathbf{i}$

(d) $y\mathbf{i} + x\mathbf{j}$

(e) $(2y, xy, -xz)$

(f) $2x + 2y$

Solutions for Appendix G

G.1 The result is the same.

References

1. Batchelor GK (1967) An Introduction to Fluid Mechanics. Cambridge University Press, Cambridge.
2. Bejan A (2004) Convection Heat Transfer. Wiley.
3. Bird RB, Stewart WE, Lightfoot EN (2007) Transport phenomena. John Wiley & Sons, New York.
4. Brodkey RS, Hershey HC (2003) Transport Phenomena: A Unified Aproach. Brodkey Publishing.
5. Bureau International des Poids et Mesures (2006) The International System of Units (SI). Organisation Intergouvernementale de la Convention du Mètre.
6. Clark MM (1996) Transport modeling for environmental engineers and scientists. John Wiley & Sons, New York.
7. Costa Novella E, et al. (1983) Ingeniería Química. Alambra Universidad, Madrid.
8. Crespo A (2006) Mecánica de fluidos. Thompson, Madrid.
9. Davis TG (2005) Moody Diagram. MATLAB Central File Exchange, <http://www.mathworks.com/matlabcentral/fileexchange/loadFile.do?objectId=7747&objectType=file>(Feb. 2, 2008).
10. Faghri A, Zhang Y (2006) Transport phenomena in multiphase systems. Elsevier, Amsterdam.
11. Fox RW, McDonald AT, Pritchard PJ (2005) Introduction to Fluid Mechanics. Wiley.
12. Hirschfelder JO, Curtis CF, Bird RB (1964) Molecular theory of gases and liquids. Wiley, New York.
13. Incropera FP, DeWitt DP, Bergman TL, Lavine AS (2006) Fundamentals of Heat and Mass Transfer. Wiley.
14. Kockmann N (2008) Transport phenomena in micro process engineering. Springer-Verlag, Berlin Heidelberg New York.
15. Kundu PK (1990) Fluid Mechanics. McGraw-Hill, New York.
16. Kays WM, Crawford ME PK (1980) Convective heat and mass transfer. Academic Press, San Diego.
17. Landau LD, Lifshitz EM (2003) Fluid mechanics. Butterworth-Heinemann, Oxford.
18. Law CK (2006) Combustion Physics. Cambridge University Press.

19. Leal LG (2007) Advanced transport phenomena – Fluid mechanics and convective transport phenomena. Cambridge University Press.
20. Reynolds WC, Perkins HC (1977) Engineering thermodynamics. McGraw-Hill, New York.
21. Schlichting H (1979) Boundary-layer theory. McGraw-Hill, New York.
22. Slattery JC (1999) Advanced transport phenomena. Cambridge University Press.
23. Warnatz J, Maas U, Dibble RW (2001) Combustion. Springer-Verlag, Berlin, Heidelberg.
24. Welty JR, Wicks CE, Wilson RE (1984) Fundamentals of momentum, heat and mass transfer. John Wiley & Sons.
25. White FM (2007) Fluid mechanics. McGraw-Hill.
26. White FM (2005) Viscous fluid flow. McGraw-Hill.
27. White FM (1991) Heat and mass transfer. Addison-Wesley.
28. Williams LG (1985) Combustion theory. Addison-Wesley.

Index

Made in the USA
Middletown, DE
08 March 2019